"*Curious?* is a soul-building read that will change forever the way you think about embracing new opportunities. Combining well-designed self-help with state-of-the-art positive psychology and profoundly inspiring stories, this is the perfect book to read when you are having second thoughts about challenging yourself to explore that next step in life!"

—Stephen Post, Ph.D., and coauthor of
Why Good Things Happen to Good People

"*Curious?* will wake you up to the rewards, adventures, and meaning inherent in both life's most significant and most quotidian moments."

—Sonja Lyubomirsky, Ph.D., author of *The How of Happiness:
A Scientific Approach to Getting the Life You Want*

"How do you know what makes you happy? How do you figure out what you really want? By acting on your curious impulses! Todd Kashdan gives us the tools, language, and plan to put a life change into effect. And the journey can be as happy as the destination."

—Marci Shimoff, author of *Happy for No Reason*

"Curiosity is what makes for great leaders, great parents, and great friends. Buttressed by the latest scientific research, *Curious?* is one of those rare books that can make you rethink how you see the world."

—Arianna Huffington

"Todd Kashdan, a pioneering scientist of positive psychology, has written a book that's more than just another take on positive thinking. *Curious?* points the way to an exploring spirit that leads to wonderment, joy, and meaning. It's one of those rare books that is both research-based and practical."

—David G. Myers, Hope College, author,
The Pursuit of Happiness

"Todd Kashdan explains in practical terms how you can develop your innate curiosity to make your life more rewarding, interesting, and meaningful."

thor of *Happiness: Unlocking the Mysteries of
h*, and the Joseph R. Smiley Distinguished
Professor at the University of Illinois

Curious?

Discover the Missing Ingredient to a Fulfilling Life

Todd Kashdan, Ph.D.

HARPER

NEW YORK · LONDON · TORONTO · SYDNEY

HARPER

A hardcover edition of this book was published in 2009 by William Morrow, an imprint of HarperCollins Publishers.

HarperCollins books may be purchased for educational, business, or sales promotional use. For information, please e-mail the Special Markets Department at SPsales@harpercollins.com.

First Harper paperback published 2010.

Designed by Ashley Halsey

The Library of Congress has catalogued the hardcover edition as follows:

Kashdan, Todd.
  Curious? : discover the missing ingredient to a fulfilling life / Todd Kashdan.—
1st ed.
      p.   cm.
  Includes bibliographical references and index.
  ISBN 978-0-06-166118-1 (hardcover)
  1. Self-realization. 2. Happiness. 3. Curiosity. I. Title.

  BF637.S4K333   2009
  158.1—dc22

                                                                2008055373

ISBN 978-0-06-166119-8 (pbk.)
HB 06.16.2023

To the five women in my life,
Sarah, my best friend and life partner
Raven and Chloe, my little beacons of curiosity and exploration
My grandmother, the fountainhead of strength and resilience
My mother, who left this world far too young to
appreciate her lasting impact

Contents

Acknowledgments

One of the benefits of writing a book is that I can give a public proclamation of gratitude to the important characters in my life.

It goes without saying that one of the reasons this book came into being is an article D. T. Max wrote about me in *The New York Times Magazine* (Sunday, January 7, 2007) about my course on the science of well-being. I received a large number of media requests and a number of invitations to write a book that describes the broad complexity of well-being, beyond happiness. It continues to astound me how seemingly arbitrary moments can change your life if you are open and willing to explore them.

With boundless support by administrators and fellow faculty members in the psychology department at George Mason University, I have been given absolute intellectual freedom with my research program. Thank you for providing an environment that allowed this project to germinate.

I also had the generous financial support of several organizations, including the National Institute of Health, the Anxiety Disorder Association of America, and the Veterans Integrated Service Network, that allowed me to study what I loved. I never served in the military and thus, it is with great honor to provide some small sign of my appreciation by treating combat veterans suffering from unbearable anxiety and producing research showing the strengths that veterans continue to possess (decades after their tours of duty). It's my hope that something in this book can aid veterans in their often difficult transition from war back to civilian life.

Science can be a lonely endeavor. Forging the kind of relationships that energize, inspire, and allow me to effortlessly be myself has been essential.

For the colleagues offering collaboration, expertise, mentorship, support, and needed input during this enduring passion of mine, thank you: Arthur Aron, Catherine Bagwell, Ingrid Brdar, Kirk Brown, Joseph Ciarrochi, R. Lorraine Collins, Ed Diener, Susan David, Jon Elhai, Robert

Emmons, William Fals-Stewart, Frank Fincham, John Forsyth, Ray Fowler, David Fresco, Barbara Fredrickson, Jeffrey Froh, Chris Frueh, Matt Gallagher, Amie Grills-Taquechel, Dóra Guðmundsdóttir, Steven Hayes, James Herbert, Stefan Hofmann, Stephen Joseph, Terri Julian, Laura King, Carl Lejuez, Jan Loney, Alex Linley, Shane Lopez, Sonja Lyubomirsky, James Maddux, Doug Mennin, Nexh Morina, John Nezlek, Nansook Park, William Pelham, Christopher Peterson, John Roberts, Paul Rose, Mark Russ, Richard Ryan, Carol Sansone, Suzanne Segerstrom, Leslie Sekerka, Martin Seligman, Ken Sheldon, Ryne Sherman, Marci Shimoff, Jerome Short, Paul Silvia, C. Rick Snyder, Charles Spielberger, Jeff Stuewig, June Tangney, Howard Tennen, James Thompson, Gittendre Uswatte, Amy Wenzel, Andy Williams, Kirk Wilson, Alex Wood, Mantak Yuen, and Michael Zvolensky.

Without the incredible students working in my Laboratory for the Study of Social Anxiety, Character Strengths, and Related Phenomena, I would be much less productive and the research process would be far less rewarding. As if running my laboratory was not enough, in the making of this book, William Breen and Patty Zorbas provided careful editing and challenging comments for the central chapters. Melissa Birnbeck assisted me in background research and Christine Plummer brought unwavering competence, commitment, and assorted help to this project. This book is demonstrably better by having them on my team.

There are others who contributed to this book by sharing their unpublished research, giving me permission to use their questionnaires, and often unbeknownst to them, providing essential ideas, knowledge, criticism, and penetrating insights via email or conversation. I especially want to thank the many friends, relatives, acquaintances, clients, students enrolled in my positive psychology and science of well-being courses, and strangers who generously contributed their stories or helped find them for this book. I couldn't possibly list everyone but look forward to the times ahead when I return these favors.

Just as I stumbled upon the topic of curiosity in a serendipitous fashion (described in this book), the same goes for some of the most important people in my life. For the invaluable conversation, encouragement, generosity, compassion, wisdom, and always welcome playful diversions during the writing of this book, I am particularly indebted to Robert Biswas-Diener, Adam Harmon, Patrick McKnight, Michael Steger, and Martin Stern. To bear the burden of my hypomania, my appreciation will endure. I hope in some capacity I offer a small percentage of what you offer me.

Thanks to my editor, Mary Ellen, and the entire team at Harper Collins for supporting me and this project before my vision evolved. Their confidence in me provided a well-needed springboard to take risks and persist during good and trying times.

Thanks to my agent, Richard Pine, for his insight, candor, optimism, intelligence, and dependability. He is one of the toughest critics I ever met and as a result, his words of encouragement often mean the most. Rarely a conversation goes by between us when I am not jotting down notes or spending valuable time contemplating his words. I couldn't imagine working with anyone else.

Then there is my "second" editor, Janet Goldstein. She was my bedrock foundation of strength and encouragement. Her editing and labors were crucial in creating the book I wanted to write. She was invaluable to this project, and in the process, I found a close friend.

Through my cousin Randy, I was able to land a job in a world of neon and debauchery: the floor of the New York Stock Exchange. Great memories. But I am equally thankful to my cousin Eric who helped me escape. He was the only member of my family who supported my ambition to be a scientist, professor, therapist, and writer. This single source of encouragement provided the extra fuel I needed to persevere and secure a lasting, renewable source of meaning in my life.

Even though he had absolutely nothing to do with this book, my life would be incomplete without my twin brother, Andrew. He regularly reminds me that with the right mindset, we can unearth a wealth of novelty in the seemingly familiar. Despite prenatal adventures together and a lifetime of shared experiences, nobody intrigues me more. I wish the same sentiments for everybody in every relationship.

Writing this book has been one of the most challenging and meaningful experiences in my life. Thankfully, the most important people in my life were with me for the entire ride. Few people get a second chance at finding loving, generous parents. With Barry and Marilyn Spitz, I did. Barry has been my wise counsel for this project and nearly every other for as long as I have known him. Marilyn encapsulates that perfect combination of boundless curiosity and tenderness to which every adult should aspire. Nobody will replicate what my grandmother, Selma, has done in terms of caring for the generations before and after without complaint. I admire her courage and humility from afar as she has them in an abundance I could never match. She remains my role model of how to live a life that matters.

As for my wife, Sarah, this book could not have been written without her. Her sacrifice is immeasurable. With the patience of a Tibetan monk, she listened to my flurry of ideas during parties, romantic dinners, moonlit strolls, and attended to random epiphanies that often occurred at ungodly hours of the day. There was no escape from this book, and she let it be. Raking the crust from her eyes, she cared for our twin girls, Chloe and Raven, when I was absent more often than I wanted to be. She remains my most intimate confidante, and perhaps the best thing about her is that she still can make me laugh absolutely anytime she wants. It is a power that continues to astound me, and one that she abuses regularly. Besides everything else, she found the perfect title for this book. My love only grows deeper as I watch Chloe and Raven flourish in her arms.

Curious?

1
Seeking a Fulfilling Life
Why There Is More to It Than Happiness

What do you want most in life? For the vast majority of us, the answer is "to be happy." We want to be happy, and we want our children and loved ones to be happy too. In a survey of more than 10,000 people from 48 countries, happiness was viewed as more important than success, intelligence, knowledge, maturity, wisdom, relationships, wealth, and meaning in life.

Not surprisingly, we spend a lot of time, effort, and money trying to achieve this sometimes elusive happiness. Our desire to "be happy" has resulted in a multibillion-dollar self-improvement industry with a staggering number of books, tapes, motivational speakers, therapists, and gurus. Infomercials plead with us to purchase the latest top-secret happiness strategy—appearing when we're most vulnerable, teetering on the edge of sleep at 2 A.M. In recent years, houses of worship have become empires, megachurches where members can pray, eat, shop, sing, and work out in a one-stop happiness enterprise. It is simply exhausting to think about the myriad of choices available to create happiness.

Despite the commercialization of its pursuit, happiness is a good thing. And though some deny this fact, it's hard to take them seriously. Yet is happiness itself truly the thing, the goal, that we should constantly be seeking? Does the pursuit of happiness even work? Are we becoming happier?

I would answer "no" to all of these questions and here's why: *We have been sold on the idea that being happy is the only or most important goal in life.*

While acknowledging the importance of happiness to creating a fulfilling life, when we focus on it, we lose out on the complexity of being human. We ignore the other important pieces of the puzzle, such as meaning in life, maturity, wisdom, and compassion.

Happiness involves feelings or judgments about life. It is a gauge we use to measure if our life is going well. But if we do things solely to be happy, what happens after this goal is met? We can find ourselves on an endless, fruitless, treadmill of "happiness seeking," never actually feeling all that happy! I believe that all the time and energy spent looking at and trying to move the happiness gauge a notch higher can be used much more effectively. Instead of constantly trying to be happy, we should focus on building a rich, meaningful life, guided by our core values and interests.

This is what the Dalai Lama means when he says that "the very purpose of life is to seek happiness." How is he defining happiness here? He isn't talking about feeling good or satisfied. He is talking about living a life infused with meaning while embracing the full range of human experiences—the positive *and* the negative.

This is not a happiness book. This is a book about being alive, exploring, discovering insights, and living a life that matters. This is a book about the science behind creating a fulfilling life, broadly defined.

You might be feeling as if you want it all. I want to be happy, I want meaning and purpose in my life, I want greater flexibility and acceptance, I want to become wiser and more mature, I want to be successful and creative, I want to be able to tolerate the distress and uncertainty of being human, and so on.

When we recognize that there is more to a fulfilling life than being happy, then we are left with an important question that this entire book hinges on. Namely, what is the central ingredient to creating a fulfilling life?

The answer is curiosity.

Reacquainting Yourself with Curiosity

On the surface, curiosity is nothing more than what we feel when we are struck by something novel. During long car rides, road signs plead with us to pull off at the next exit to visit underground caverns and carnival rides. Your carload of travelers may pass by these opportunities with mixed responses—teenagers squealing with intrigue, Grandma detesting anything screwing up her window vista, and your own hesitation to spend money to satisfy childish interests. You might experience similar moments of beckoning when an odd book title compels you to see what's inside or when you are enthralled as you pass the rising gull-wing doors of a parked DeLorean. Commonly, we view curiosity as what we experience when opportunities bump into us.

Curiosity does draw our attention to things that are interesting. But while this appears simple, my research and that of others shows that curiosity is a deeper, more complex phenomenon that plays a critical role in the pursuit of a meaningful life.

We often make the fatal mistake of thinking growth opportunities come to an end when something or someone becomes part of our daily routine. When things become familiar and predictable, we become mindless drones. We tune out. As soon as we think we understand something, we stop paying attention.

Novelty is different. We often pay attention to the unfamiliar and listen to new people because they grab our attention. What we forget is that novelty always exists in the present moment. There is much to learn from the unfamiliar *and* the familiar. No two hugs are the same, no two pizzerias make pizza slices the same way, no two times we watch *The Godfather* are the same, and so it goes. Being curious is about recognizing novelty and seizing the pleasures and meaning that they offer us.

Another reason that curiosity is neglected is that it operates below the surface of our desires. It's not as simple as thinking positive, being optimistic, being grateful, being kind, or feeling good. Being curious is about how we relate to our thoughts and feelings. *It's not about whether we pay attention but how we pay attention to what is happening in the present.*

The past and future are nothing more than symbolic representations of reality. Mental photographs (past) and dreams (future). Real events happened, but once the action is over, they are a figment of reality. These representations aren't even stable; they change. You look at pictures in your photo albums and re-create memories. We create stories around the narratives we want to weave about ourselves and other people. Was Theodore Roosevelt the last of a dying breed of candid, authentic presidents who could swim naked in the pond behind the White House in broad daylight, kill hundreds of animals with his shotgun, fight police corruption, and be the brave champion of the environment who created the national park system? Or was he a sociopath with a grandiose vision of his own importance, driven by uncontrollable impulsive urges, who found a good outlet to hide it? History is rewritten constantly and the past changes.

Our moods change constantly and thus our ideas about the past change with them. As for the future, it remains unwritten. Anything can happen, and often we are wrong. The best we can do with the future is prepare and savor the possibilities of what can be done in the present.

Only in the present can we be liberated to do whatever it is we want. It's a razor-thin moment when we are truly free. When we are curious, we exploit these moments by being there, sensitive to what is happening regardless of how it diverges from what it looked like before (past) or what we expect it to be (future). We are engaged and alive to what is occurring. *We are energized. We are open and receptive to finding opportunities, making discoveries, and adding to the meaning in our life.* To reiterate, it's not about being attentive; it's about the quality of our attention.

By being open and curious in our moments, we can improve even the most mundane aspects of our daily routine. How is it that we walk down a supermarket aisle and remain oblivious to the hundreds of items on the shelves? One reason is that they don't interest us. We've been in the same store a million times, we're harried, and we might hate grocery shopping in general. If we're on a mission to make a Caesar salad with the aid of a written shopping list, our attention is laser-focused on locating the items we need. The owners of the supermarket try to outwit us with colorful product displays that force their way into our consciousness. Without help to navigate through the flood of objects in our inner and outer worlds, chaos would reign. This is where curiosity steps in, if we let it.

If we are open to unplanned opportunities, we intentionally seek out interesting items in the store. We may find a new food to get our picky kids to eat, a delicacy to spice up our week, or maybe we waste a few dollars on a culinary experiment that doesn't agree with us. Regardless, there is something enjoyable about being unsure of what to expect and being willing to seek out discoveries and act on opportunities as they arise.

Shaping our Abilities and Identities

Besides transforming trivial moments, curiosity offers a gateway to creating profound intimacy, insights, and meaning in life. The best way to grasp curiosity is to witness it. Consider Stacy. I was introduced to her through a friend who described her as a kid masquerading as a top-notch architect. I wanted to learn a bit more about her passion, so we met on a plot of land where a huge, expensive house was being built near the Blue Ridge Mountains in Virginia. She had been hired as a consultant on the project.

I watched her as she looked, listened, and hiked a long way out into the forest. I asked what grabbed her attention. "The land, and the way it slopes. What you could do to capitalize on the view. Where the house should go. How to keep from cutting down too many trees."

Every so often, she posed her own questions. "What are most of these trees, what kind, I mean? Do you know how deep the dirt is? Can you drill a well in this ground? After all, there's lots of shale and limestone." Without being asked, she shared her overall impressions.

"I think a house should have a relationship with the place it sits, don't you? The house should share a vocabulary with the environment. I mean, think about all the people who lived around here and how they lived. When they built houses, they built them in a certain style for certain reasons and used what's local. Now we know about designs from all over the world, but should we use them here?"

When you look at the world through Stacy's eyes, you can see how curiosity shapes our identities and the story of our lives. It's more than feeling good in the moment, it's more than taking advantage of novelty in front of us, it's an attitude toward living.

Sylvan Tomkins, a pioneer social psychologist, is perhaps the first researcher to explore the unique value of positive emotions, 35 years before the "positive psychology" movement came along. He believed curiosity was of profound evolutionary significance:

> The importance of curiosity to thought and memory are so extensive that the absence . . . would jeopardize intellectual development no less than the destruction of brain tissue . . . there is no human competence which can be achieved in the absence of a sustaining interest.

Since he made this statement, much work has accumulated to demonstrate the truth of this working hypothesis. If we are to find fulfillment more readily and sustain it, we need to harness our curiosity. In the words of Tomkins, "I am, above all, what excites me." Just how this mechanism works and how we can more fully understand and harness its benefits is made possible by the emergence of a secure, scientific foundation.

Curiosity is hard-wired in the brain, and its specific function is to urge us to explore, discover, and grow. It is the engine of our evolving self. Without curiosity, we are unable to sustain our attention, we avoid risks, we abort challenging tasks, we compromise our intellectual development, we fail to achieve competencies and strengths, we limit our ability to form relationships with other people, and essentially, stagnate.

When you look at successful people in history, it becomes clear that curiosity motivated them to expand their horizons, and it has to be given at least partial credit for their accomplishments. Whether it is scientists, artists, architects, politicians, CEOs, owners of mom-and-pop shops, advertising executives, teachers, political activists, or nearly anyone trying to discover something, their stories highlight the powerful motivator of curiosity in why we do the things we do.

Curiosity is part of the architecture needed for building a pleasurable, engaging, and meaningful life. Understanding this motivation allows us to better understand human nature.

Why is curiosity so important? First, just like Stacy's quests, curiosity motivates us. This is captured by our greater interest in and preference for new experiences and information. When we experience curiosity we are willing to leave the familiar and routine and take risks, even if it makes us feel anxious and uncomfortable. Those who deal better with novelty, who function more optimally in a world that is unpredictable, uncertain, and unstable, I call "curious explorers." Curious explorers are comfortable with the risks of taking on challenges. In fact, the most curious among us actually lust for the new.

Second, curiosity helps us extract and integrate experiences and meaning from these new experiences. When we are probing the nooks and crannies of new worlds, inevitably we are also learning. This is the ultimate goal of our curiosity system: to add to our existing knowledge, skills, and competencies. These additions help us to better understand ourselves and the outside world, cope with the challenges of everyday life, and improve our ability to handle chaos.

Resisting our opposing craving for certainty, we discover that the greatest rewards come when we question authority, question the status quo, question our beliefs, and question everything. Sometimes these rewarding, positive feelings come from anticipating the excitement of exploring something that we might not want to end. Other times, it's only after getting the experience we craved that we feel excited, satisfied, or content. We have only one life to experiment with, and pulling off our blinders can help us get started.

The Thieves of Childhood Curiosity

Children are born with boundless curiosity. In the beginning, infants get excited by brightly colored objects, slight changes in the sound of their mother's voice, and anything within their small field of vision. Without any shortage of fascination, hours can be devoted to watching the spinning blades of a fan. As their senses become more developed with age, their wonder and excitement expands, and they venture out a bit further. Every cupboard and drawer hides a mystery worthy of investigation. At birthdays and on holidays, boxes and gift wrap are as interesting as the gifts themselves. Children seek a deeper understanding of the complex world around them and play offers a safe opportunity for them to make some of the fearful unknown known.

But something happens to this innocent and courageous exploration. Society gets in the way. We are given an endless series of rules and obligations that keeps our curiosity in check. *Do it now, ask questions later. Stay away from strangers. Avoid controversial topics and hot-button issues. Your elders are always right.* Instead of learning about history, science, and great literature, we are taught how to do well on standardized tests. Memorizing facts too often replaces the intrigue of being exposed to foreign, exotic territory—the sensation of touching the pouch of a mother kangaroo, imagining a time when dinosaurs roamed the earth, scrutinizing the merit of improbable ideas such as reincarnation and cryogenic immortality.

Rules are useful because they provide structure and help us avoid doubt and uncertainty. But they get in the way of freedom. Following directions conserves energy, but following one's unique direction expands energy.

Rather than being encouraged to learn about ourselves and our interests, we are more often taught how to make decisions about what to do with our lives as early as possible so we won't waste time achieving our goals. Pick an academic major, choose a career, and start a family. Whether other interests are squelched isn't important. What's important is to "make something of yourself," "be able to support yourself," and "realize that life is more than just having fun."

We also live in a climate of fear. Headline news reports bombard us with warnings of what we should be wary of. Terrorism, kidnappings, credit-card and Internet scams, mad cow disease, sexual predators, plane crashes, killer bees, shark attacks, tire recalls, and more. As a result, we do everything in our power to stay out of harm's way.

But there's a risk to playing it safe. Our actions are dictated by what we don't want instead of what we do. Aspirations are put on hold. But the things we fear are often unlikely to happen. We fear a terrorist attack, a child kidnapping, a mad gunman at school—yet the chances of any of these things happening is infinitesimally small.

Most people overestimate risk, failure, and danger and underestimate the value of being curious. It is time to reclaim our neglected strengths. We can—and should—choose how we want to live our lives. Are we governed by fear and the need for safety, or are we willing to accept a bit of risk and anxiety in the pursuit of satisfaction, growth, and meaning?

Living a life of curiosity is not about ignoring risk and anxiety. It's about being willing to do what one values, even in the face of risk and anxiety. What if we saw things as they really are without judgments of what we expect them or want them to be? What if we acted on our curiosity when deciding what to do with our free time, what careers to choose, who to spend our time with and devote our lives to?

Curiosity serves as a gateway to what we value and cherish most. We can reclaim the lost pleasures of uncertainty, discovery, and play from our youth. Outside of childhood, our innate curiosity does not come as easily or naturally, but these obstacles can be acknowledged, confronted, and ultimately, overcome. We can cultivate curiosity to shape our lives closer toward the direction of where we want it to be. Realizing our potential depends on it.

To be sure, curiosity is not the only quality that contributes to a happy, meaningful, and fulfilling life. That being said, it's hard to think of a human endeavor where curiosity doesn't play a vital role. Think about it: science, technology, work, business, sports, education, politics, arts and aesthetics, introspection, personality development, relationships— you name it. Few areas are more worthy of our time and effort than enhancing our ability to be curious.

This is not an unsupported claim. As you will see, the ideas in this book and my conclusions are supported by a large body of scientific work, much of it known only to readers of academic journals and little of it integrated into a meaningful picture relevant to our everyday lives. This book will provide tools you can use to renew the relationship with the curious explorer within you.

Our Ability to Get Easily Accustomed to Positive and Negative Changes

There is a reason to be skeptical of curiosity and any new strategies that might lead to lasting fulfillment. We possess an amazing ability to rapidly readjust our moods and feelings in response to changes in our circumstances. Scientists refer to this as our well-being "set-point." Except for a few major unpleasant events such as becoming a widow or being stricken by a serious disability, disease, or unemployment, we quickly become familiar with changes in our life and drift back to how we felt beforehand. This point we return to after bad or good things happen to us is our typical mood. This typical mood is genetically determined, relatively out of our control, and somewhat immune to what happens to us except for brief downward and upward shifts.

This is a fantastic resource because it prevents us from being overwhelmed by life's challenges and tragedies. Among other unpleasant events, you are going to be rejected, you are going to fail at something important, and you will get ill and injured more than once. It's wonderful to know that we overcome most of the obstacles thrown our way and bounce back.

Human beings generally want to predict, understand, and control what happens to them. After all, if we didn't try to gain closure about an event, we would be in an eternal state of limbo, incapable of making decisions or drawing conclusions about what happens to us. This might seem like a good thing after an unpleasant event. After all, being able to gain closure gives us a sense of comfort. We feel secure and safe; we feel confident; we feel intelligent.

But what happens to us after a positive event? When your romantic partner kisses you, saying, "I want to spend the rest of my life with you," you will probably experience loving feelings. Among other pleasant events, you are going to feel grateful when you experience someone's kindness, learn something new, accomplish difficult feats, and hug, kiss, and make people laugh. Settling on a clear explanation and coming to closure about what a positive event means to us, dampens our positive feelings and cuts them short. The parts of our brain that are activated by new experiences, the pleasures of savoring the moment, the connections between nerve cells that are created when we try to make sense of what is going on, come to a screeching halt. The positive experience is over as soon as we label it, filing it away in our library of prior knowledge.

So the big question lingers. Is it possible to shift our set-point upward and sustain increases in our well-being? Yes, and the answer lies in curiosity. By themselves, simply being optimistic and hopeful will succumb to adaptation. The same goes for efforts to increase other strengths such as kindness, forgiveness, humor, bravery, or zest. That's why I am not giving you another book about thinking positively or being friendlier.

If we want to prolong positive experiences, if we want a more fulfilling existence, we need a new mindset. Acting on our curiosity is the alternative to seeking closure and certainty. What gets in the way of our ability to adapt and return to our regular routine is the same thing we try so hard to get rid of: novelty and uncertainty. As long as something is novel, we are still in the process of finding and creating meaning. When something is uncertain, there are multiple possible outcomes for what can happen next. The ending is unclear and thus, the event is still ongoing.

Curiosity is different than other ways of being fulfilled in that it's about appreciating and seeking out the new. It's about being flexible, recognizing the novelty and freshness of the familiar. Instead of trying desperately to explain and control our world, as a curious explorer we embrace uncertainty. Instead of trying to be certain and confident, we see our lives as an enjoyable quest to discover, learn, and grow. There is nothing to solve, there is no battle waged within us to avoid the tension of being unsure. We don't take positive events for granted, we investigate and explore them further. Recent scientific findings show that when we are open to new experiences, when we are flexible in how we approach people and tasks, when we relish how the unknown far outweighs what we know, positive events linger longer and we extract more pleasure and meaning from our world. This book is about unleashing the curious explorer within you.

The Yin and Yang of a Fulfilling Life:
A New Approach with Curious Exploration
as the Engine

As a clinical psychologist, I treated scores of children and adults suffering from debilitating mental health problems. Several of my clients felt life wasn't worth living. I will never forget the people who reached a point in their lives where suicide was considered a viable option; they haunt me to this day. I learned a great deal about what it means to live fully and completely from working with them. The well of despair some patients showed in therapy was certainly tied to their suffering. They wanted the pain to go away and saw suicide as the ultimate escape from the self. But suffering is not the cause of suicidal thoughts. People try to kill themselves because of the absence of the positive. They can't find a reason to live. In their minds, their life lacks meaning and purpose.

Discovering this was a turning point in my thinking. My ability to understand human nature had been strangled by a fanaticism to describe and catalog people's problems while ignoring their positive experiences and strengths. By ignoring the positive and focusing on the negative, like many therapists who devote their lives to helping people, I was dismissing half of what it means to be human.

When I contacted clients a year after treatment ended, I discovered that many showed few if any signs of the problems that had once rendered them helpless. Yet, far too many seemed unfulfilled and lost. Many of my former clients were living lives that were little more than a mechanical routine—wake up, work, sit in front of a television or computer screen, go to sleep, and repeat—with a few scattered moments of deep engagement such as births, weddings, and deaths. I became acutely aware that it was not enough to focus solely on what psychologists call the "presenting problem." Adding positive experiences with the goal of balancing the pull of the negative in a person's life is a starting point.

But focusing on the positive is only one part of the equation. I needed a more complex model that reflected the actual lives of the people I knew and the people I treated. In the often one-sided focus on positive experiences and strengths, positive psychology and happiness researchers underestimate the value of negative feelings, uncertainty, and stress. Negative thoughts and feelings cannot be ignored or avoided. Effectively handling the pain and stress that life inevitably brings is an essential part of creating a rich, meaningful existence.

If you search for a tensionless life, you will find little to be satisfying. Coping with negative experiences and being psychologically flexible are often springboards to peak experiences and personal growth. Throughout this book, we'll look at how this works.

Both the positive and the negative need to be addressed. Anything less fails to capture the entire package of what it means to be human and lead a satisfying, fulfilling life.

What You Will Discover in This Book

We'll start by examining the foundations of curiosity and what our life can look like if we build and wield it. In Chapter 2, we will explore the scientific evidence that supports the need to invest in curiosity. By the end of the chapter, you will understand the nature of curiosity, the benefits linked to curiosity, and how curiosity is unique from other positive qualities and human strengths. You will become aware that even if you are not a very curious person now, it is a strength that we can develop. In Chapter 3, we will adopt an evolutionary perspective to understand the origins of curiosity. Then, we'll explore the latest brain science to show that curiosity recruits brain regions that are distinct from those that "light up" when we just feel good. Our brains respond with more intense, longer-lasting pleasure when we are exploring and growing. At the same time, you will learn that there is more than one type of curiosity. From curiosity that is accompanied by joy and excitement to curiosity that is best described as an intense craving to know in hopes of relieving the tension bubbling inside us. There is nothing pleasurable about being curious about that rattling noise in your car or wondering whether the red patch on your cheek is a zit, goiter, or cancerous sore. You'll learn that you might not always find pleasure, but you will build knowledge, skills, strengths, or meaning, and perhaps all of the above.

The next part of the book plays a decisive role in showing you how you can apply these ideas to your own circumstances and enhance your well-being profile. Chapter 4 shows you how to invest your most limited and valuable resources—time and energy. This investment, when intentionally focused, can radically enhance life's moments and events to be more interesting and engaging. You will become better at creating interest even in mundane, trite, and appallingly boring situations. Curiosity doesn't occur by magical happenstance or self-talk; it is set in motion by emotional qualities and specific behaviors. You will gain insight into several hands-on strategies and the science behind how and why they work. In Chapter 5, we move from the story of how to understand and harness the "curious moment" to how these moments can be deepened, strengthened, and transformed into lasting interests and passions. Once again, you will be presented with a set of detailed strategies and learn how to move from understanding the science to applying it to your own life. This chapter also addresses ways that you can apply the lessons for sustaining interest to inspire and motivate others.

Although many books have been written about the importance of relationships to a happy life, at the core of Chapter 6 is a new perspective about relationships and social connections as the ultimate source of personal growth. At the fore, I will show the logic of attending to what we don't know about people and how they differ from us, instead of relying on what we do know about people and how they are similar to us. We will begin with the art of sparking interest among potential friends, lovers, and colleagues; move on to the challenges of creating deep, meaningful connections over time; and confront head-on the issues of boredom and stagnation that far too often creep into existing relationships. Because there is no single secret to improving our social world, a number of scientifically informed guidelines and strategies will point the way to maintaining passion and excitement in long-term relationships.

Any temptation to ignore the negative and focus only on the positive is dispelled in Chapter 7.

Anxiety is as powerful a motivator as curiosity. It can be beneficial, annoying, or debilitating. I will show you how instead of being the cure, common techniques used to control anxiety are often the biggest problem of all. Do not skip this chapter out of the mistaken assumption that you are not pained by anxious thoughts and feelings. Each of us is hardwired to experience this common emotion. I will explain how we can change our relationship with anxiety and by doing so, open space for greater opportunities to detect and create meaning. You will be exposed to research by me and others showing that becoming a curious explorer can provide relief to problematic anxiety.

Despite the profound benefits of curiosity, there are times when curiosity can be meddlesome, bizarre, counterproductive, and even sinister. Chapter 8 is a look into the dark side of curiosity. There are people who show intense interests in sex, death, and other areas that run counter to being a healthy, useful contributor to society.

The book ends with a vital chapter on discovering meaning and purpose in life. By the end of Chapter 9, you will be equipped with knowledge of how passions at the center of our identities, otherwise known as a purpose or "calling," develop. Contrary to popular belief, there is no evidence to suggest that we have a preordained purpose in life. Instead, based on the latest evidence, concrete guidelines for increasing opportunities for creating meaning and purpose are presented. Having a purpose will be shown to be one of the paths to a healthy, satisfying, and long-lasting life. It is not the only path, but it is an important one. You will also learn how being a curious explorer can help find meaning and benefits in those difficult times when beliefs about the world are disrupted by loss and trauma. Even if you are fortunate enough to avoid these hardships, this chapter is worth reading. An arsenal of insights will leave you better prepared to extract meaning when things go right and when things go wrong. Curiosity is at the heart of resilience and recovery.

Taken together, my aim in writing this book is to affect you in two ways. First, by showing you how and why curious exploration is integral to a well-lived life. Second, by providing clear strategies for you to become a curious explorer and to extract greater pleasure and meaning from all of life's moments and to invest in lasting passions and fulfillment.

All of the information in this book draws on the scientific evidence, much of which comes from my own work and that of other scientists whom I was grateful enough to work with or learn from. It is a culmination of 10 years of research on the role that curiosity and novelty play in well-being, the development of meaningful relationships with other people, and transforming feelings of anxiety. It also includes the insights from my work as a clinical psychologist working directly with people struggling with anxiety and depression, trauma, and addictions. I have spent most my professional life thinking about what it means to have a fulfilling life. This book is about exactly that: how to become a curious explorer and how to build a life infused with meaning. I hope at some point after reading this book when someone asks you what you want most from your life, instead of quickly replying, "All I care about is being happy," you might respond, "I want to be curious and do something that matters."

2 The Curiosity Advantage
Opening the Gateway to What Makes
Life Most Worth Living

Most people stop looking when they find the proverbial needle in the haystack. I would continue looking to see if there were other needles.

Albert Einstein

When asked about his uniqueness, Albert Einstein didn't blabber about his intelligence, work ethic, happiness, or relationships—he talked about his curiosity. He claimed that his accomplishments had to do with an appreciation of the little mysteries of everyday life that others often take for granted. There was always something he didn't understand that he relished trying to figure out. His drive to search, ask questions, and explore the vast unknown was as important to him as the drive to find answers. It is an approach to living that is simple to understand and rare in practice.

Even though we may not develop our natural curiosity with Einstein-like devotion, when we act on our curiosity, we feed our brains and are in the greatest position to enrich our lives. Forty-five-year-old championship dog trainer Ben is a case in point.

Ben was a successful trial lawyer. At an early age, he became the judge of an appellate court for military veterans. Listening to the appeals of soldiers who served our country and felt justice failed to serve them, Ben made an impact. However, he experienced a quiet but gnawing lack of connection to what he was doing. The stress and emotional intensity of his work led him to feel less rather than more engaged with his professional life. When I met him, he had been exploring various interests for several years. He took piano lessons and Pilates classes. Then he bought a dog. Little did he know that buying this dog, a Tervuren named Vixen, would be a turning point that would open up new roads for filling his days, adding new layers to his personality.

The breeder helped Ben train Vixen, and his dog became a better companion. During one of these increasingly frequent training sessions, the breeder mentioned that Tervs have a reputation as stellar sheep herders, sharing the name of a farmer who teaches sheep herding. On a whim, Ben called and signed up for weekly lessons. A far cry from the excitement of courtroom testimonies and surprise witnesses, I asked Ben why he likes sheep herding so much.

"For one thing, the man who teaches me is wonderful. I'm in awe of his connection to the dogs—he's the Dog Whisperer. As for me, when I'm training Vixen, my brain is on overdrive, attuned to how my emotions and body communicate certain messages to him. At the same time, I have to be aware of what Vixen's trying to tell me. And all the while, I have to be aware of what's going on around us, making adjustments, using my body language in a way that is graceful and unnoticeable as possible to the judges."

He added, "If you had told me that I was going to fight my way to become a judge only to spend most of my salary on sheep-herding lessons, I would have thought you were nuts! But here I am!"

His work with Vixen is far removed from the strains and challenges in the courtroom. Ben started waking up invigorated, and his new world started to spill over into his "real" job. Other people noticed his new enthusiasm. Whether it's due to Vixen, his receptiveness to new experiences, or some combination, Ben's life is changing, his personality is evolving, and he's better for it.

Curiosity Is the Engine of Growth

If we are aware of and receptive to our curious nature, we can catch serendipity. Told he would herd sheep as an adult, a younger version of Ben would have laughed. It's all too easy to pass on intriguing possibilities that fall outside of our comfort zone. We put off engaging with the new for a future day. But what are we waiting for? Burn out? Retirement? But why wait for big changes to add intriguing elements to your lifestyle? Why wait for burnout? Why wait until retirement?

Ben shows us how curiosity is always available to us and something we can act on at any time. When we are interested in anything—friends, shopping, picking out a recipe, or looking up the definition of a new word—curiosity is at work. When we actively inject variety into our ordinary routine, curiosity is at play. Small, innocuous changes—from mixing cereals from four boxes in our cabinet to changing our jogging route—can enhance our days and lives. These repeated experiences modify our brain, further enhancing our ability to flourish in a world with more unknowns than knowns.

But curiosity is not just an effortless and spontaneous experience. It is an intense motivation woven into the basic fabric of our personalities. All of us, to varying degrees, are driven to seek out new and uncertain experiences. Although you might believe that certainty and control over your circumstances brings you pleasure, it is often uncertainty and challenge that actually bring you the most profound and longest-lasting benefits.

Curiosity is about recognizing and reaping the rewards of embracing the uncertain, the unknown, and the new. There is a simple story line for how curiosity is the engine of growth.

By being curious, we explore.

By exploring, we discover.

When this is satisfying, we are more likely to repeat it.

By repeating it, we develop competence and mastery.

By developing competence and mastery, our knowledge and skills grow.

As our knowledge and skills grow, we stretch and expand who we are and what our life is about.

By dealing with novelty, we become more experienced and intelligent, and infuse our lives with meaning.

Curiosity begets more curiosity because the more we know, the more details that we attend to, the more we realize what there is to learn. Why? When we embrace the unknown, our perspective changes, and we begin to recognize gaps—literal and figurative—that weren't apparent before.

The physics of rock climbing provides a vivid illustration. Beginning rock-climbers rely on their instincts. Their goal is not to fall. They cling close to the face of the mountain, and use a lot of upper body strength to pull themselves up. After bagging a few peaks, they start to accumulate some counterintuitive tricks. For instance, to avoid slipping or falling, you push your body away from the mountain. This might seem like a sure-fire way to fall, but what happens is that you can get a larger surface area of your toes and feet on the mountain. You have a more stable grip, and you are less likely to fall.

As rock-climbers get more comfortable with their hand and foot positions, they start to see places to move their feet and grab with their hands that they didn't see before. Their fingers get stronger and instead of just looking for pieces of rock that jut out, they start looking for cracks to stick their fingers and toes into. Their expertise allows them to see more gaps and be more creative as a rock climber.

Curiosity isn't always a joyful process of discovery, and it's not just an upward spiral to some hoped-for peak experience. Sometimes curiosity is an itch that we need to scratch. Rather than being motivated to explore because of a sense of wonder or intrigue, we crave specific answers to a puzzle.

We crave specific information that might help us close the gap—and this is what brings pleasure. For instance, an adult needs to figure out the last few answers of a crossword puzzle; a teenager has to learn about Franz Kafka for a writing assignment; a manager wants to figure out the strengths of her workers to maximize productivity. There are also questions loaded with palpable tension, from the mundane to the cosmic: Am I at risk for Parkinson's or Huntington's disease like my parents? Is my marriage going to last?

Sometimes joy is central to our explorations and other times, the sole purpose is to release tension and find meaning. In either case, finding answers to specific explorations is not the end of our curiosity. The more knowledgeable or skillful we become, the easier it is for us to recognize other holes in what we know and other questions and interests we want to explore.

Curiosity is at work whenever we feel open and receptive to experiences that offer more than what is already known. Each time we embrace the new, we can't help but extend or expand ourselves. It could be as simple as listening to a previously unheard musical chord on the radio. It could be as profound as reflecting on whether the person sweating and panting beside us in bed is just another sexual partner or someone we love—and discovering that we are in love.

Uncertainty Intensifies and Prolongs Pleasures—and You Don't Even Know It

Asked about his daily routine in a televised interview, Deepak Chopra, a leading writer and teacher on spirituality and consciousness, extolled the virtues of curiosity:

> I wake up with the hope this day is even more uncertain than yesterday.
>
> It's the unknown that we live, breathe, and move in all the time thinking it is the known.
>
> If a life can be a series of perpetual surprises, that's the most joyous experience you can have.

Most of us have not learned this lesson. We don't realize that trying to make sense of the unknown actually adds to the intensity of our positive experiences and makes them last longer than pleasures gained from familiar routines.

The supportive evidence is considerable. In one experiment, thirsty research subjects had their brains scanned while drinking water (pleasurable) or Kool-Aid (more pleasurable). Some subjects were told what they would be drinking and when (certain) and others were left in the dark (uncertain). Brain circuits linked to positive feelings lit up like a street carnival when subjects knew they were getting a drink but were left wondering what drink was going to be served. Just like a busload of gamblers spending eight hours wed to a slot machine at a casino, excited about their prospects of winning every time they pull the lever, being uncertain enhances the pleasure of positive events.

Our expectations influence how we are going to feel, and good news is more enjoyable when it comes as a surprise. Olympic athletes content to be competing who don't expect to win anything have been shown by scientists to be more likely to gush with joy after winning a silver medal than do those who were expecting to win.

Becoming fulfilled is about much more than the size or type of gifts that life presents to us. There is something special and meaningful about being surprised, uncertain, and unsure of exactly what is going to happen and how things will turn out.

Interestingly, we are often clueless about what will bring us the strongest, longest-lasting pleasure and meaning in life. The psychologist Timothy Wilson was interested in why it is particularly meaningful when strangers are kind to us. Is it the gift or the unexpected kindness that touches us most? (After all, we are taught at a young age to be wary of strangers.) Perhaps "random" acts of kindness *are* more meaningful. When we don't expect and can't explain the reason for someone's kindness, the mystery deepens our pleasure and changes the way we view the world (maybe people *aren't* so annoying). These moments of surprise offer opportunities for "meaning making."

Wilson's research team strolled through a college campus giving students cards with a silver dollar taped to it. All the cards were signed by "The Smile Society," but some students' cards included details on what this odd society does, which is brighten people's days. Students who didn't learn why they received a free dollar felt better about the gift, and these feelings lasted longer. Being mystified enhanced their positive feelings.

In another study, instead of being given a card with a silver dollar, subjects were asked to imagine how they would feel if they were the recipient of a silver dollar. Half were told they would learn who and why the person gave them money; the other half were told they would only be given the stranger's name. Asked to imagine how they would react, people told who gave them the gift and why expected to be in a more intense, enduring positive mood. The ones who were left wondering why they were given a gift expected to be disappointed. Students thought that understanding what happened to them was essential to feeling good.

These findings are fascinating because they suggest that what we think brings us pleasure and meaning is often *the exact opposite* of what does. Our "mispredictions" have tangible repercussions. If we believe that understanding everything, being able to confidently predict the future, and being in control are necessary, then we are going to drift toward stagnation. Doing things that are only mildly pleasurable, we will underestimate two profound sources of happiness and meaning in life: novelty and uncertainty.

Curiosity Is a Counterbalance to Certainty, Closure, and Confidence

On the surface, curiosity and the need to be certain are both about gathering information and reflecting on experiences. Dig deeper and you discover some important differences in how we relate to our world, other people, and ourselves. When we are curious, we are open to expanding ourselves and to do so, we explore. When we seek certainty, we are looking for finality. We are looking to close the search process sooner rather than later so we can feel confident that we know what to do and how to act.

To do so, we interrogate without the openness and awareness that characterizes the curious explorer. We can see the difference in the contrasts below:

Curiosity creates possibilities; the need for certainty narrows them.

Curiosity creates energy; the need for certainty depletes.

Curiosity results in exploration; the need for certainty creates closure.

Curiosity creates movement; the need for certainty is about replaying events.

Curiosity creates relationships; the need for certainty creates defensiveness.

Curiosity is about discovery; the need for certainty is about being right.

From the onset, I need to be clear that none of us should be embarrassed or blame ourselves if we possess the need to be certain. Our environment often pushes us toward a narrow and certain view, say in the situation of excessive noise (where we can't think clearly) or time pressure (where we are overly focused on finishing rather than doing well). In these overstimulating environments we are closed off, trying to survive instead of thrive.

Other people and society also give us a clear message that there are benefits to being certain and decisive, to finding firm answers to questions and closure, and, when possible, to avoiding ambiguity and the tension that it brings. Isn't this what you want from your favorite news anchor and journalist when there is talk of how to invest your money, plan for the hurricane season, or remain healthy and prevent life-threatening diseases? Isn't this what you want from your auto mechanic when you ask about the rattling noise in your car? You feel safe and secure when they can pinpoint the exact problem.

Unfortunately, there are costs to working hard to feel safe, secure, and confident. We often end up shutting down our search for information too early in the process. In essence, we quickly become close-minded. We protect our beliefs even when they might be wrong. If someone shares our views, we praise them; if someone questions or challenges our views, we criticize, attack, and discount them.

Lack of curiosity is a breeding ground for:

- Stereotyping and discrimination, that in the extreme leads to hatred and even violence

- Inflated confidence and ignorance that leads to poor decisions
- Dogmatism and rigid thinking, which is the opposite of psychological flexibility

We need to be wary of the need for certainty. Seeking certitude can cause our beliefs and decision making to crystallize prematurely, and the resulting reluctance to consider new information can hurt us in the long run. From the research on this topic, we know the sad irony that the greater a person's need for certainty, the more confident they are that their ideas are "right."

Would you want a romantic partner who shuts down and refuses to see you as anyone other than the person they met on those first few dates? Would you vote for a politician who refuses to remain open to new knowledge and developments that would improve their precision and wisdom? The same goes for parents, physicians, scientists, jury members, and company managers. If one thing is certain, it's that things are always changing. Fail to accept this truth, fail to be open to and curious about the new, and stagnation is assured and often, far worse.

Curiosity Liberates Us

A student of mine went to her first hockey game with her boyfriend. She was aghast at seeing her quiet, reserved lover yelling at referees, arguing with other fans, and jumping out of his seat to hurl profanity-laden tirades at the players. The experience really shook her up, and when she asked him about it, he laughed and said, "Honey, I just had my game face on! It's how I let off steam after a hard day." She smiled and realized there remains much to appreciate and discover about her new partner. Never to be fully solved, a relationship—like many important things in our lives—can be viewed as an exciting puzzle, the pieces of which we continually discover.

How many of us arrive home from work to carefully attend to our roommates, children, pets, and homes as if we are seeing them for the first time? As a scientist, therapist, teacher, husband, and father, it has become clear to me that most of us fail to appreciate their scent, their touch, and how they contribute to our lives in ways that change from day to day.

Curiosity motivates us to be open to viewing the world, other people, and ourselves from multiple perspectives. These angles might conflict with how we once viewed things or how others view things. We might need to change our belief systems to accommodate them. It can be difficult to modify deeply held beliefs that previously gave us a sense of certainty and clarity. Yet, when we are in a curious state of mind, we can tolerate this tension.

Curiosity Inspires Attention: Wherever You Go, if There Is Mindfulness, There Is Curiosity

Each of us experiences moments of interest every day that we can either ignore or explore. Whether we view our curious states and feelings as openness to experience, novelty-seeking, flow, intrinsic motivation, or a search for meaning (all terms used by researchers, psychologists, and spiritual leaders), curiosity is at their core. They all involve feeling open and receptive to experiences that offer more than what is already known.

One other incarnation of curiosity is especially important to explore because it has a 2,500-year history as a window and path to well-being. It is the curiosity that comes from mindfulness. From the earliest Buddhists to generations of philosophers, to more recent scientific investigations, mindfulness has been found to have profound benefits.

In short: Pain is unavoidable; suffering is optional. Mindfulness paves the way to face stress and hardships in a way that minimizes suffering and maximizes fulfillment.

Unfortunately, it is a challenge to be mindful. Our default mode of thinking is mindlessness. We reflexively label and categorize things. For instance, other people get categorized as our friends and perhaps inner circle, or they are outsiders whom we tolerate, ignore, or shun. We form lasting impressions in mere seconds about whether something is ugly or beautiful, or whether tasks are boring or interesting. We don't let ourselves feel anxious, sad, or angry, without labeling these feelings as desirable or undesirable, good or bad. Depending on our judgment, we decide to accept and work with our feelings, or we struggle to avoid or change them.

We are also constantly comparing ourselves to standards: my legs are flabby wind chimes compared to everyone else here who is wearing shorts (I am not going to the beach today); how can someone say something so stupid (glad I'm not them); my parents would definitely not approve of this woman if I brought her home (should I end this romance before it comes to that?); if I'm not funny nobody is going to like me; and so on. Although it would be nice if our brains would stop their constant chattering and shut up once in a while, this inner dialogue is part of being human.

Although some people are less likely to rush to label and categorize everything, we all do it. Processing information this way is efficient, and we are hardwired to do it. Humans are creatures of habit, and it requires effort to be reflective and avoid labels. That being said, the research is clear that the more we automatically and mindlessly categorize thoughts, feelings, and other people, the more we suffer. Well-being stumbles when we go on autopilot.

If you look closely at mindfulness, you will learn what I have learned. Mindfulness cannot exist without curiosity. Reclaiming your own curiosity will allow you to truly understand what mindfulness is, and it will improve your ability to reach this exalted state.

Mindfulness is a way to relate to ourselves and the world around us. In speaking of mindfulness, scholars and practitioners mention being fully aware, living in the present moment, being nonjudgmental, accepting, compassionate, and open. This last part, being open, is what often gets left behind in the rush to describe it as succinctly as possible. I think this omission is a huge mistake. Consider some of the seminal and recent definitions:

- "Keeping one's consciousness alive to the present reality"
  —Thich Nhat Hahn
- "Paying attention in a particular way: on purpose, in the present moment, and nonjudgmentally"   —Jon Kabat-Zinn

- "Awareness without judgment of what is via direct and immediate experience"   —*Oxford English Dictionary*

On first glance, it seems that mindfulness is about how we use our attention. But it's so much more than intentionally paying attention to ongoing moments and being aware of what happens in the present.

As a kid, I remember leafing through a 50-cent book called *Bankei Zen* at a garage sale. The book described a state of mind called the "unborn Buddha mind." Curiosity is what Buddhist scholars are talking about when they refer to the "unborn Buddha mind" or "beginner's mind." *Bankei Zen* was talking about seeing things with fresh, open eyes as if for the first time. Just like a child who sees a turtle for the first time, crawling in and out of the water with its stubby legs paddling away, we take an interest in our inner and outer worlds with the same receptivity and wonder. This is mindfulness.

Without curiosity, discussions about mindfulness are about nothing more than paying attention and trying to concentrate—on purpose. When your mind is closed, you are preoccupied with the past or future. Examples that look like mindfulness but aren't include:

- Tired, driving late at night, focusing intense effort to stay in your lane. Instead of being mindful, you are laser focused on the task of driving safely.
- Listening to a customer talk while repeatedly reminding yourself to give them their receipt. Instead of being mindful, you are pushing through a work assignment.
- Looming over a child as they peel an apple with a knife. Instead of being mindful, you are poised for danger to ensure their safety.

In these instances, you are concentrating and resistant to change. You fuse with the situation. Concentration is useful, especially in the situations described, but at a cost of lost moments. Mindfulness is about being guided as opposed to controlled by planning. Concentration is the blunt power of directing attention and holding it steady. Mindfulness is delicate and sensitive, bringing energy, creativity, and a more expansive space to play in.

In 2004 a number of scholars met and reached a consensus about how to define mindfulness. Finally, the word *curiosity* was introduced as one of two powerful, interlocked components.

- The first has to do with the self-regulation of our attention. It involves our ability to continuously shift and refocus attention on inner experiences or the outer world in the present moment. It involves monitoring and sustaining attention in a flexible manner to the target of our senses.
- The second has to do with the quality of our attention. It has to do with the adoption of a specific type of orientation toward immediate experiences and is characterized by curiosity, openness, and acceptance. When we are mindful, we maintain an inquisitive nature about the thoughts, feelings, behaviors, and events experienced in the present. This inquisitive nature allows us to be receptive and aware of all aspects of reality.

Mindfulness is about moments. When taking a shower, you are fully aware of the sensation of water hitting and rolling down your body. Standing in the shower today is different from any other episode in our lives because of the emotions and thoughts we carry, the placement of our bodies, the tenderness and tension in various body parts, and so on. When your mind bursts forth with a new idea, you can purposely guide attention to it and then gently return to the physical feeling of each dose of water. Without openness to experience and flexible attitudes, mindfulness doesn't exist.

Take Shane, a bridge toll booth collector, as an example of learning to be mindful. At work, Shane's contact with the outside world is limited. For every smile directed at him, dozens of drivers pull up to his booth and look at him with contempt, shout obscenities at him to hurry up, or try to fly through without paying. After several years on the job, his body began responding to the chronic stress. He developed physical ailments, including a partially paralyzed stomach, which caused him digestive trouble. He was treated by a doctor who suggested an unusual intervention. He gave

Shane a plan for short-circuiting the tension before it built up and caused so much pain. The plan involved a systematic approach to paying attention. As Shane described,

> Throughout the first week I worked a lot on just realizing how my body feels. In the morning, at work, and each night before I go to bed, I kind of do a body check. Each time I sit down in my booth, or reach over and talk to a customer, I do the body check. I've always kind of done this at the gym, but now I am doing it for my entire body. Being summer, I was sweatin' like a bastard out there in the booth, and it had a great effect on what each muscle of my body felt like. I'd say to myself, my calves feel relaxed, my thighs feel solid as an oak tree, my knees hurt, my arms feel like cinder blocks. I'm a sitting time bomb. So at first all I did was feel worse. I'd think, well that's great, Shane, now you know that your arm feels like you just towed a fire truck with it. After a few days, I started realizing that my body and emotions matched so much more than they did before. For instance, if I was stressed, my shoulders killed me; if I was nervous, my body shook a little; if I was joyous, my face felt like I put a warm washcloth on it; if I was content, my legs melted into the chair I was in like hot wax. This made me feel whole. It was like I wasn't separate pieces, but rather pieces making up a whole.

As illustrated by Shane, we can do more than pay attention; we can take an interest in what's right in front of us or inside us and explore. We can learn new things and find unexpected rewards. Curiosity provides the calm energy that allows mindfulness to unfold.

Curiosity Is Part of Our Personality, All of Which Is Clay that Can Be Molded

Moments of curiosity are the building blocks to becoming a curious person. Even though curiosity is hardwired in all of us, we differ in how we are oriented to the world and how we process information. Some people are more open, and others are more resistant to change and uncertainty. Curiosity is an essential part of our identities, and we all fall somewhere on a continuum. That is, it's not a question of whether or not you are a curious person; rather, it's the degree of curiosity that you possess. To understand where you reside, it's useful to break down the qualities that characterize a very curious person (the high end of the continuum).

The first quality is **intensity**. Highly curious people feel more intense feelings of curiosity, interest, and wonder. They show a strong preference for novelty, variety, and complexity. The more curious among us are open to new experiences, even when familiar and secure ideas and routines are challenged.

The second quality is **frequency**. Highly curious people feel curious many times in a given day, and the urge to explore and investigate is easily triggered. Brief sightings, sounds, and tidbits of information fuel their curiosity with the greatest of ease. Feeling the breeze of a motor scooter zip by leads to a flurry of questions about how fast they go, what kind of gas mileage they get, and whether there are fewer fatalities and lower maintenance costs than cars, and wonder about why they are the most popular transportation in urban parts of Europe and Asia but rare in the United States.

The third quality is **durability**. A curious person's sense of intrigue and desire to explore often endures for lengthier periods of times. Think of the curious sports fan. This person goes nuts for games that are tied and go into overtime. Even though all games start with pure uncertainty, the high-curiosity fan is disappointed and leaves early when one team demolishes the other. The person with a high-curiosity profile wants to prolong the uncertainty as long as possible because the tension is exciting. For curious people, the pleasures of novelty and uncertainty linger longer and are purposely extended.

The fourth and fifth qualities are best understood together as **breadth** and **depth**. Breadth refers to the number of events that make a person feel curious or interested. Curious people are interested in a large span of their world such as their work, relationships, passions and hobbies, and their own private thoughts, feelings, and values. We can be curious about the past, present, or future; we can be curious about the external world, other people, and ourselves (introspection is an act, and the curious person enjoys figuring themselves out; the less curious person cares less about examining the inner motives that guide their choices). Once a person finds something that captures their interest, depth refers to whether they hold on to that interest and integrate new experiences into their identity. That is, initial interests can lead to long-lasting hobbies or passions, providing a renewable source of excitement and meaning.

When we are very curious, our lives are densely packed with moments of being open to exploring whatever offerings are the targets of our attention. We can change the frequency of curious moments in our life as well as how curious we are as a person. The latest scientific evidence is clear that repeated experiences can change our brains and are the first signs that our personality is changing. This is what scientists call "neural plasticity," and it is a major breakthrough. Until the 1990s, it was assumed that any opportunity to modify our personalities came to an abrupt halt in early adulthood. We now know that our personalities are malleable and, in particular, that we can become more (or less) curious.

On one hand, scientists have shown that being open to new experiences, taking an interest in learning new things, and being able to adapt to new situations decline substantially in middle-to-late adulthood. This finding echoes the stereotypical, closed-minded older adult deriving little pleasure or meaning from what interested the playful, more carefree, younger versions of themselves. On average, curiosity and openness to experience increases (for both boys and girls) from age 12 until they attend college. By the mere age of 30, the curious explorer inside them starts fading away. Social responsibilities, likes and dislikes, and life paths start to solidify as we close ourselves off to newness. Younger adults might fear this slow deterioration of curiosity more than death itself (listen to The Who belting out "My Generation"). I see this attitude played out in the comments of my students and the older adults with whom I work in therapy ("I should have seen you a long time ago before *I became who I am*").

But stereotype is not destiny, and it wasn't the ACLU, AARP, or politically correct public service announcements that fought the concept of "you're never too old to change." It was science.

In the last 10 years, a number of scientific teams began reporting their results after following the same people for decades to uncover what parts of our personalities remain consistent over the course of our lives and what changes. We are talking about scientists observing children until they enter college. In other cases, scientists tracked subjects for four decades beginning at age 20! From this laborious undertaking, we now know that our personalities, including how curious we are, can change across our life spans. For instance, out of 870 intellectually gifted kids who were observed from ages 12 to 16: Over these pivotal four years,

- 5% showed a decline in curiosity
- 44% showed an increase in curiosity
- 51% remained the same.

To put these findings into perspective, 49% of kids changed in their curiosity. When you look at the other "big" personality traits, such as being gregarious, agreeable, or impulsive, only 34%–38% of kids changed. A hallmark feature of prepubescent kids is their difficulties in understanding and handling their negative feelings. Yet, only 43% showed a change in their anxiety, embarrassment, and temper tantrums. What this means is that of the major, universal personality traits of human beings, curiosity and openness are the most malleable.

Because the playful, boundless curiosity of childhood doesn't show a slow, steady, systematic decline for everyone, it raises an important question. What can we do to be in that bracket of people who actually increase (or at least retain) their curiosity beyond age 30? This book is devoted to providing the answer to that question along with concrete strategies for doing so.

Based on several studies, we have some tentative answers about the type of behaviors at age 20 and 40 that predict openness, curiosity, and the ability to handle and even embrace uncertainty at age 60 (a full 20 or

40 years later!). Those college-aged youngsters identified as very curious in later adulthood had several characteristics in common:

- They had rich emotional lives filled with both positive and negative feelings (further evidence that negative emotions are not "bad").
- They were actively searching for meaning in life (this included questioning authority and dominant, widely accepted ideas).
- They didn't experience themselves as being restricted by social norms.
- They chose careers that gave them opportunities to be genuine, authentic, independent, and creative.

Each of these qualities suggests ways to retain the intense curiosity and preference for the novel (over the familiar) that characterizes enlightened youth. They also point to the opposite. A hyperfocus on seeking security, avoiding distress, and sticking to a comfortable routine lessens one's curiosity and, in turn, satisfaction and meaning in life.

Contrary to popular belief, our personality can change even if we are no longer in our awkward adolescent years. We can't change our genetic code, gender, age, ethnicity, or prior life events (the good, bad, and ugly). But we can change how we think and which activities and goals we invest in. Even if you feel little connection to these qualities now, you can become a curious explorer. Science gives us reasons to be optimistic.

When we speak about the possibility of change, we have to be concrete about how this happens in practice. We know that emotions last for seconds, moods last in the vicinity of hours, and our personality is something that is forged over the years. If we want to change, we need to begin our work with instantaneous events, practicing the mindset of being a curious explorer. This will help change our brain in subtle ways, which will eventually weave new strands into our personality. As Tibetan Buddhists remind us, "If we take care of the minutes and moments, the hours and days will take care of themselves."

The "Big Five" Benefits of Harnessing Curiosity

We can see from many vantage points that curiosity provides the motivation to explore and grow as a person. Curiosity is a centerpiece of mindfulness, and contrary to what we often expect, novel and uncertain events intensify and prolong positive experiences. Even though each of us is endowed with a certain level (something we can see even in infants), with effort, we can become more curious and in so doing, we can change how we approach the world.

The benefits of curiosity affect all the nooks and crannies of our lives, but I like to talk about what I see as the "big five" scientific discoveries on the benefits of being curious. They are big because they are not controversial. Each is a desirable advantage, no matter where you live, what your life situation is, or what your goals are.

Health

It's hard to argue for something more useful than averting death for a while longer. More than 2,000 older adults aged 60 to 86 were carefully observed over a five-year period and those who were more curious at the beginning of the study were more likely to be alive at the end of the study, even after taking into account age, whether they smoked, the presence of cancer or cardiovascular disease, and all the rest of the usual markers. It is possible that declining curiosity is an initial sign of neurological illness and declining health. Supporting this idea, patients with Parkinson's and Alzheimer's disease suffer from a degeneration in dopamine circuits and, in turn, show a lack of interest in novelty and a general unwillingness to explore their environment. The same goes for older adults with degeneration in the brain regions linked to curiosity. Why is this important? Because with scientific advances in health and nutrition, we are living longer but not necessarily better. An unanticipated price of longevity is the estimation that one of every eight members of the baby boomer generation will develop Alzheimer's disease. After age 65, the risk of Alzheimer's disease doubles, and by 2030, the number of Americans with this disease is predicted to grow by 50% (from 5.2 to 7.7 million). Nonetheless, there are promising signs that enhancing curiosity reduces the risk for these diseases and even the potential to reverse some of the natural degeneration that occurs.

We see the same results in the animal world. Female rats who frequently seek out new experiences (fiddling around with toys in their cages, exploring wings of mazes) live longer than their more cautious peers. Using a strain of laboratory rats that all eventually die from tumors, more curious rats lived 25% longer than those fearing the unknown. Now if we find anything even remotely like this in humans, we need to start taking curiosity seriously and investing in ways to harness it.

Intelligence

If intelligence is about analyzing problems and solving them, what could be better than being able to handle the novelty and uncertainty that creeps into everyday decisions? Supporting this idea, using a sample of 1,795 three-year-old children, curiosity and intelligence were measured at age 3 and again eight years later, when the kids turned 11. What the researchers found is that even after you take into account how intelligent kids are at age 3, being more curious led to greater intelligence over the course of eight years. Even more amazing is that highly curious kids scored 12 points higher on IQ tests compared to less curious kids. To put this into perspective, the average person has an IQ of 100, doing better than half of all scorers. An increase of 12 points moves them up so that only 22% of all scorers do better! Quite impressive. Thus, even when you account for a child's initial intelligence, if they also happen to be highly curious, they show intense cognitive development in their formative years.

In adults, higher curiosity is regularly tied to greater analytic ability, problem-solving skills, and overall intelligence. Together, high curiosity and high intelligence is the powerhouse combination that characterizes the best students, workers, managers, scientists, artists, and other luminaries who contribute so much to the people and world around them.

The idea of improving people's intelligence is appealing. Yet, attempts to enhance intelligence such as with the Head Start Program (early education program for infants and toddlers) more often than not have been colossal failures. It is unclear exactly how resistant intelligence is to improvement, but a century of research suggests that approximately 50% of our intelligence can be explained by our genetic blueprint and another 10% reflects both how we were raised and whatever happened during our prenatal womb adventures.

While intelligence is quite resistant to change, curiosity can be cultivated, and it is available to anyone who desires a fulfilling life. Many prodigies in a wide variety of disciplines start off by playing music, writing poetry, or shooting hoops because it feels good, but this pleasure is often obliterated by the pressure to succeed. Basically, when curiosity and interest disappears, the benefits go with them.

When we remain hyperfocused on intelligence, we miss out on an important truism: The most intelligent people are not necessarily the best and brightest. Curiosity predicts better grades and achievement test scores in school, and curiosity in the classroom is a better predictor of students' willingness to transfer knowledge learned into long-term interests and careers. Even mentally retarded children who are curious show an intense motivation and dedication to learning that rivals their "higher-functioning" peers. We need to ask whether the goals of teaching children is for them to do well on tests and get into the best colleges or for them to be successful. If it is the latter, the cultural obsession with intelligence needs to be reconsidered and curiosity needs to be brought to the forefront.

Meaning and Purpose in Life

My own work has shown that curious people possess an optimal orientation to ensuring the continual building of a fulfilling life. They are immersed in the present, feeling as if their life is infused with meaning, and simultaneously, they are receptive to searching and finding meaning and adding it to their existing supply.

How do we know this? In several studies, we asked people to track what they do each day and to what degree they extracted meaning from these activities for several weeks. With this approach, we discovered what people actually did in a given day and week, as opposed to relying on retrospective reports of what they thought they did. This is important because we know that if you ask parents of young children if their lives are enjoyable, they invariably provide an enthusiastic "yes!" to questions. Yet, if you ask them to report on what happens during the course of a particular day, you get drastically different results. Parents report extreme frustration and unhappiness more frequently than moments of joy. How you reflect on your past is often dramatically different than what actually happened in the moment.

Curiosity is also the entry point to many of life's greatest sources of meaning. We can refer to them as interests, hobbies, and passions. If we are going to find our calling and passions in life, it is going to stem from something that unleashes our curiosity and fascination. Exploring and discovering is the trial-and-error process of following up on things that are interesting and determining which are worthy of investment for the long haul. Curiosity is the driving force behind finding and creating small and sustainable pockets of meaning. It could be spending time with a significant other or the experience that comes from complete immersion in a conversation, book, movie, game, or sport and then making sense of the event. This process will be discussed at length in a later chapter of this book.

Social Relationships

We don't often equate curiosity with social relationships. Yet, I firmly believe you cannot form and maintain satisfying, significant relationships without an attitude of openness and curiosity. One of the top reasons that couples seek counseling or therapy is pure boredom. This is often the starting point of resentment, hostility, communication breakdowns, and a lack of interest in spending time together (which only adds to the initial problem). Curious people report more satisfying relationships and marriages. Happy couples describe their partners as interested and responsive.

Besides existing relationships, curious people act in certain ways with strangers that allow relationships to develop more easily. In one of my studies, people spent five minutes with a stranger of the opposite sex in a conversation to get acquainted, and then each partner made judgments about their partner's personality. We also separately questioned people's closest friends and parents to gain insight into the impressions that curious people formed on the people who know them best. Acquaintances of a mere five minutes, close friends, and parents all view highly curious people as highly enthusiastic and energetic, talkative, interesting in what they say and do, displaying a wide range of interests, confident, humorous, less likely to express insecurities, and lacking in timidity and anxiety compared with less curious people. Curious people ask questions and take an interest in learning about partners, and intentionally try to keep interactions interesting and playful. This interest contributes to more satisfying and meaningful interactions that, in turn, often lead to the development of relationships.

In the absence of curiosity and openness to experiences, people show an intolerance of uncertainty and a strong need for closure in their lives. While these characteristics might aid in protecting a person from anxiety and stress, their destructive influences on social relationships are far ranging. Less curious people rely on stereotypes to describe others and find new information that is inconsistent with these beliefs to be threatening. As a result, they cling even more strongly to their first impressions even when they are wrong. This closed-mindedness is the springboard of prejudice and the rapid rejection of those who disagree or fail to conform.

To avoid the discomfort of uncertainty in social relationships, less curious people quickly shift from love to hate, trust to mistrust, and other forms of black-and-white thinking. Curious people are comfortable working through doubts and mixed emotions in their relationships. Less curious people view ambivalence about a relationship partner as a sign of unworkable problems that quickly escalates into extreme action: violent disagreements, rejection and distrust, and abrupt break-ups. The origin of societal ills can often be traced to the absence of curiosity.

Happiness

For many, this is the big kahuna, the fundamental aim of life. So it's worth ending this scientifically guided advertisement for curiosity with a discussion of how it contributes to our happiness.

More than 130,000 people from more than 130 nations, representing 96% of the globe's population, were recently surveyed by the Gallup Organization. Scientific surveys have never witnessed anything like this before. The two factors that had the strongest influence on how much enjoyment a person experienced on a given day were "being able to count on someone for help" and "learned something yesterday." What this Gallup poll confirms is that developing good relationships with other people, seeking out the new, and growing as a person are keys to a "happy" life.

That's not all. In one of the largest undertakings in the field of psychology, Martin Seligman and Chris Peterson devised a scientific classification of the basic strengths that human beings can possess. This system was the end result of reading the works of ancient philosophers, religious texts, and contemporary literature, identifying patterns, and finally subjecting these ideas to rigorous scientific tests. Their research led to a final tally of 24 strengths. Some of them, such as love, spirituality, and emotional intelligence, are the first that come to mind when we think about what will make us happy (perhaps you have read books about these claims). When Seligman and Peterson evaluated the 24 strengths, curiosity was one of the five most highly associated with

- Experiencing overall life fulfillment and happiness
- Taking satisfaction from one's work
- Living a pleasurable life
- Living an engaging life
- Living a meaningful life

Curiosity was more important than other highly touted qualities such as love, spirituality, emotional intelligence, kindness, forgiveness, perse-

verance, wisdom, and others (beaten out by only hope, zest, and gratitude). In case you're wondering, these studies are based on the answers to
website questionnaires by samples of 4,000 to more than 12,000 people
from countries around the globe. For more than 1,000 young adults from
the United States and Japan (aged 18–24), at the height of discovering
their identities, the only strengths with stronger links to happiness than
curiosity were hope and zest.

In a survey of 839 Croatian students, my colleagues and I found that
of the 24 strengths, *only curiosity and zest were consistently part of the top five
strengths linked to attaining a pleasurable, engaging, and meaningful life!*

It's a bit silly putting strengths into a competition where 24 enter and
only a few emerge as winners. When we look at the top strengths that
predict happiness, each of them often appears together with curiosity.
As an example, in one study, about 1,200 teenagers wore a watch that
beeped them eight times per day for a week. The teenagers carried
around a packet of questions with them, and they wrote down what they
were doing, thinking, and feeling at the time they were beeped. It's a
killer approach to studying a day in the life of curious kids. The 207 kids
(the top 10%) who were most interested in what they were doing
throughout the week were labeled "curious explorers"; the 207 least interested (the bottom 10%) were labeled "bored kids." How did curious
explorers differ when they were beeped as they went through their day?
They felt more optimistic and hopeful, confident, and believed they
were in control of their actions and decisions as opposed to feeling
governed by what others think. Instead of feeling like "pawns," with no
control over their destinies (as bored kids did), curious explorers felt a
sense of self-determination—a powerful motivator to take the reins of
life's offerings and challenges.

These findings are no more than correlations, but they hint at a bigger story. I think, like many others before me, under the right conditions
people who relentlessly pursue "personal growth" are best suited to thrive.
Enjoying the process and the outcome, and feeling like the masters of
their own destinies, curious people are in a great position to create and
sustain happiness and more.

Curiosity as Nature's Conduit
for a Successful Life

Pull down a recent book on happiness or some other similar topic from the shelves, open up the index, and look for the words *curiosity*, *interest*, and the rest of the family (*novelty*, *openness*, etc.). Maybe curiosity gets a sentence, but often there isn't a single word mentioned in hundreds of pages. As I have learned in my more than 10 years studying this topic, we cannot afford to take curiosity for granted if we want to move closer to the life we wish to lead.

Curiosity is not bound by age, gender, race, intellect, or culture. It's a universal quality of humans and the majority of our animal friends. However, we are each endowed with a temperament that makes us a bit different in our exploratory tendencies and the psychological barriers that block our way. All of this is visible both in our actions and in studies of the three-pound blob of neurons between our ears.

As a better vantage point to capitalize on and cultivate curiosity, we need to consider its origins. In the next brief chapter, we will take an evolutionary perspective and peek inside the brain.

3

Our Brains Lust for the New

The priority of our hunter-gatherer ancestors was to look out for anything that might be dangerous or deadly. Everything and everyone was evaluated for how much trouble they might cause. If I eat those plants, am I going to get sick and die? Are those howling sounds distant enough for me to safely venture out for food? Is the man in the back of the cave with breath smelling of charred human flesh a threat to me?

In the prehistoric African Serengeti, which had looming predators, injury, illness, and starvation, our ancestors realized that their chances of survival improved if they were part of a social group. It's much less threatening to attack a woolly mammoth with 10 assistants than single-handedly. As part of a group, our ancestors obtained easier access to food, care, protection, and enough sexual invitations to ensure survival and the passage of one's genes.

With these juicy benefits, the human mind evolved to fear the possibility of being rejected by the group.

The Push for the New in the Face of Rejection

Our brains are hardwired for curiosity—along with its neural twin, worry. Starting with the plight of our ancestors from 500,000 years ago, social groups always had to be selective. For instance, every cave dwelling had a maximum capacity. Groups were more willing to invest their time, energy, and resources if a new person appeared to be attractive, intelligent, competent, strong, and cooperative. Every human being was vulnerable because they were constantly being evaluated by the group. This was not the place to slack off, as the lazy and weak were the first to be tossed out to fend for themselves against hungry packs of raptors and wolves. To be accepted, our ancestors needed to prove that they could pull their own weight.

How did we protect ourselves against rejection? We constantly monitored where we stood from the perspective of other people in the group. *Do I contribute enough? Am I contributing as much as the others? Am I doing anything foolish to warrant rejection and banishment?* Over the course of thousands of years, human brains evolved to handle the job of hunting, gathering, and other survival challenges.

Unfortunately, many of the mental tools that aided the survival of our hunter-gatherer ancestors get in the way of our ability to enjoy ourselves in today's modern world. For example, prehistoric men and women who worried a lot were more likely to survive than their carefree, positive-thinking peers. Thinking negatively served as an early warning system. It triggered the brain to recognize actual and potential threats in the moment, and it also aided the brain in imagining dangerous scenarios that didn't exist. If people were prepared at all times, they were more apt to survive. Even if they were wrong 90% of the time and spent a lot of time worrying over nothing, it didn't matter because survival was the primary goal.

Negativity Bias and the Positivity Offset

We are hardwired to prepare for the worst. When danger is found—or just anticipated—our minds respond quickly with anxious feelings and a compelling desire to escape. Just as our ancestors responded with anxiety to the possibility of danger, we respond with the same intense fight-or-flight reaction when we rush to get to work on time or are about to pay for lunch when we realize that we left our wallet at home.

We spend a lot of time worrying about potentially threatening events that are unlikely to happen and, even if they do, are rarely as bad as we think they will be. Just as our ancestors worried about being banished from social groups, we worry about what other people think of us, and we feel anxious and embarrassed when we might be evaluated poorly. Ensuring that we bathe in self-loathing, our minds also do a nice job of finding and comparing us to people who are stronger, smarter, more attractive, and better equipped to deal with life's hassles (at least that's we think).

Knowing the evolutionary design of our brains helps us understand what it takes to create a fulfilling life. We are constantly trying to balance the positive and negative. Our hardwiring to attend to the negative actually has a name. It's called the *negativity bias*. Think of our ancestors hearing footsteps behind them or the rustling of tree leaves. Their hearts would beat a little faster and they would flip their heads around and get ready to fight or flee. We are designed to react quickly to potential danger. But more than that, we have the blessed curse of being able to dream up all sorts of scenarios in which we can fail, be hurt, and die, which naturally leads to even more anxiety and worrying. The negativity bias is our intense reaction to potential threat. If you are an anxious person, this bias extends further so that ambiguous and uncertain situations are experienced as dangerous until proven otherwise.

There are times when we get a break from a state of anxious alertness. Moments when there is absolutely no danger and we feel safe. In this state, we don't try to hide from the world. We can relax and rest.

This neutral, inactive state is often short-lived. There is another important element of our evolutionary design called the *positivity offset*. What this means is that when we feel safe, we show a slight bias to explore new things and seek out new experiences. We are pulled toward rewards and excitement. Without this offset, we would never learn, stretch, grow, or evolve.

Every Damn Thing Had to Be Discovered by Our Ancestors

In some capacity, every human is trying to make sense of the world, and this has strong evolutionary roots. Whether acquiring resources (tools or shelter), information (who to trust in their small tribe), or tangible rewards (food or sexual partners), our ancestors needed seeking mechanisms.

It was only when our ancestors ventured off beyond the boundaries of what was known that they could add to their knowledge and skills. An increase in knowledge and skills could help someone survive another day. Of course, just because something was new didn't mean it helped in the prediction and understanding of anything important. Remember, they had to discover everything. Fingering someone's belly button doesn't do much more than annoy the owner, rocks don't float, and so forth. It was often a creative trial-and-error process. Imagine the number of objects our ancestors discarded before they tamed fire. I suppose it's "possible" that a lucky caveman picked up flint on the first try. More likely, generations played with tree branches, bones, rocks, and anything else they could lift. They probably tried to eat, lick, carry, rub, squeeze, hit, shake, toss, dance, and chant with these objects while waiting for something interesting to happen. Once a fire started, they had to learn the connection between what did and didn't work (cause and effect).

In an unforgiving environment of animals fighting with each other to live another day, there is much to learn. To be motivated to seek out the new, our ancestors needed a hard-wired system that would cause them to eagerly anticipate and want rewards. This euphoria felt at the possibility of being rewarded compelled them to explore, seek, and investigate. After all, it's the anticipation of being rewarded that motivates and the consumption that satisfies. This anticipation and seeking is all about being curious; it is evolutionary adaptation at its best.

The Social Benefits of Being Curious

That's not all. Unarguably, curious explorers were at an advantage in gaining social acceptance compared with their less-daring peers. Their new skills would have been more attractive to their tribe and in turn would have gained greater access to the group's food, water, tools, and sexual partners. With more resources, the explorers survived, and with access to plenty of sexual partners, were more likely to pass on these traits to their offspring. It was essential to learn which behaviors made them attractive enough to be accepted by a tribe (e.g., providing food) and which behaviors ensured rejection (e.g., stealing, being lazy). Over generations, curiosity, exploration, and similar traits would be even more common.

Another behavior aiding the development of relationships is play. At the core of play is variety, whether it is devising funny nicknames for friends and teasing them, dressing up, climbing trees and buildings, roughhousing, or competing in sports and games. Play is curiosity and joy in a blender. Besides being pleasurable, play is a training ground for young children and animals to develop essential social and problem-solving skills that last a lifetime. In the animal kingdom, squirrels choose regular playmates that are outside of their family circle. Even when squirrels are separated from their childhood playmates for months and years later, when they reunite and that playmate is being attacked by a predator, squirrels are more likely to throw themselves into the mix and risk their lives to protect them. Those joyful experiences in childhood create lasting friendships that come in handy during times of need in the future. Humans show the same pattern. Play doesn't exist without a curiosity system, and regular play often leads to long-standing relationships.

For our ancestors, play, an outgrowth of the curiosity-seeking system, was one of the tools for securing allies and sexual partners, thus increasing the likelihood of survival and reproductive success. In play, we experiment with different social behaviors, rules, and roles. We expand our toolbox of how to relate to other people and what we learn about ourselves. Play is a training ground for change.

The Dance of Curiosity and Fear

Of course, not all curious, playful explorers had a happy ending. The unknown offers tremendous insights and rewards, but the explorer doesn't always survive to share it. Neighbors sat on their asses during all of these adventures, concerned about being comfortable, secure, certain, and in control; compared with the explorers, they occupied their time with the familiar and routine. They would have benefited by listening to and watching the explorers. Secondhand learning and growth is often as good as getting it firsthand, with the added benefit of not having to risk injury or death.

With the goal of being able to effectively navigate the environment, an optimal profile of flexibility emerged. The most successful people felt anxious in the face of clear and present danger and felt curious in the face of potential rewards and learning opportunities. When there was a blend of danger and reward, our ancestors experienced a conflict between wanting to escape and survive versus exploring and earning all the prizes that came with it—which might help them survive in the future.

Predictions about what might happen in the future were imperfect and, thus, our ancestors were plagued with feelings of uncertainty about what to do. We don't spend our lives cowering in fear, thirst, and hunger because our brains evolved so that even if we feel mild to moderate discomfort, most of us prefer to explore. We are hardwired to experience a rush of excitement when something novel and unpredictable breaks through the routine. Alone at the library, we peek at the bare ankle of the person next to us only to notice a tattoo with the lyrics to our favorite song.

When we are in unknown territory, leaning toward exploring instead of avoiding is the key to a rich, meaningful life; for our ancestors, it was about increasing the chances of life longevity and reproductive success. Too much or too little concern about feeling anxious gets in the way of acting on our curiosity. Absent fear, we may go beyond the limits of what's safe for ourselves and those around us. Absent of real danger, when we can't act on our curiosity, we are left with painful regrets of what our lives might have been. Our curiosity and threat detection systems evolved together, and they function to ensure optimal decisions are made in an unpredictable, uncertain world.

The 3-Pound Curious Explorer

If the motivational force of curiosity did evolve, then we should see deeply ingrained markers in our modern-day brains. With the technology that exists today, our brains can be seen for what they are—archaeological maps into our shared human history.

In the popular imagination, our brains represent the truth. People get really worked up about research showing that whether you are in love, optimistic, religious, happy, obsessive, or fearful, it all shows up in the brain. It's easy to believe that your experiences aren't valid unless they can be traced to your biology. But do you need to go into a brain scanner to find out if you love your children or whether you are miserable or pessimistic? If so, I would argue that you need to seriously consider spending less time hanging out at your local pub and more time figuring out who you are.

This being said, brain science can shed light on the origin of our curious natures and whether in fact people get more enjoyment from stability, familiarity, and certainty, or if I am right, that the key to intense, lasting fulfillment is seeking novelty and embracing uncertainty.

To begin, let's start with the most commonly discussed and misunderstood chemical in the brain that is linked to happiness—dopamine. Dopamine is a neurotransmitter that carries information from one nerve ending to another in the brain, preparing our body for action. There is an intense debate in the field of neuroscience as to what exactly triggers dopamine and, when it is unleashed, what action it inspires.

High-school biology teachers and many scientists refer to the brain regions housing the largest collection of dopamine receptors (holding docks) as the "pleasure center" of the brain. This pleasure center is the nucleus accumbens, which is part of the ventral striatum, which has early evolutionary roots in the primitive, reptilian part of our brain. The striatum connects us to what is happening in the outside world and when a reward opportunity is seen, dopamine surges through our brain.

Modern technology allows researchers to chart active brain regions as people are thinking, making decisions, and planning behavior. When volunteers undergoing a brain scan are given tasty food and drinks through a straw, erotic videos, or video games with a chance to win cash and prizes, the striatum and dopamine are stimulated. Thus, in some capacity, dopamine activity is tied to rewards. When we feel good, our neural system is flooded by dopamine. When we laugh hysterically during a movie, dopamine surges through our veins. The consensus has been that dopamine prepares our body to seize rewards and, in turn, we then feel pleasure, we smile, and we're happy.

Case closed? Not so fast. In the last few years, skeptical neuroscientists working independently found new results that challenge this interpretation—among them Jaak Panksepp at Bowling Green University, Kent Berridge at the University of Michigan, Brian Knutson at Stanford University, and Richard Depue at Cornell University. To discover how the brain generates emotional experiences, scientists often have to tweak it. In one study, Berridge gave sweet-tasting foods to a pack of rats. You can tell when rats are enjoying themselves by their facial expressions. Similar to an average 20-year-old male at the Playboy mansion, their tongues hang out of their mouths. What was different about these rats is that they were given a neurotoxin that eliminated dopamine from their brains. It was an air-tight study because without dopamine those rats shouldn't have been able to enjoy anything. The first thing that happened was that the rats stopped exploring and moving around their cages. Their food remained untouched, so the experimenters assumed that the rats lost the desire to eat and were going on a hunger strike. The experimenters force-fed them to see what would happen. To their surprise, rat tongues were soon wagging out of their mouths as they relished the sweet-tasting food. Clearly, getting rid of dopamine did not stop their ability to enjoy themselves. But it did stop them from wanting food or, more accurately, from wanting to do anything! The rats looked like a pack of apathetic teenagers zoned out in front of a TV screen.

Vermin are a nice set of captive recruits to begin understanding the brain, but eventually you have to study human beings. Besides being able to map which parts of the brain are active, we can determine whether the activity occurs during, before, or after being given a task to do. Resting in a scanner and being told that you are going to be shown pictures of an important person from your past, what parts of your brain will light up *before you see the pictures*? During the period of uncertainty before you see the pictures, will different brain regions light up compared to when you learn who it is? The mystery of who you are going to see is separate from the pleasures of looking at pictures of them and reminiscing. What goes on when we anticipate seeing someone is different from when we spend time with someone. Timing matters when it comes to how we feel and what it motivates us to do next.

By using scanning technology, we can test competing ideas about the exact role of dopamine. If dopamine is crucial to how we feel when we receive *rewards*, the brain sites where dopamine are housed should light up whether we are eating chocolate, snorting cocaine, laughing with friends, or answering questions correctly. However, if dopamine is crucial to *seeking rewards*, the same brain sites should light up when we eagerly anticipate being rewarded as opposed to the moment when we satisfy our thirst or reach orgasm. The lingering issue, which we take up next, is whether it is the pleasure itself or the wanting and craving that gets dopamine firing wildly.

Curiosity Patterns in the Brain: Seeking versus Having Rewards

Ask any parent if they think the distinction between seeking and having is arbitrary. I interviewed Mary and her daughter Fran. Fran is a typical teenager who won't get out of bed until you yell. On the cusp of flunking out, Fran tells her that if she gets her act together, she will buy her a new mountain bike with a titanium shell, shock absorbers, and other top-notch accessories. What do you think happened when she learned she might get a bike (when the "wanting" kicked in)? Feelings of gratitude littered conversations, household chores were completed, and a study group met on a weekly basis, as she radiated positivity. What do you think happened when she got the bike? She rode day and night, enjoying off-road trails in pure bliss. And at the same time, her grades slipped until by the time a month rolled around, she once again became the same person we met at the start of the saga.

When we get rewards, we are satiated. Just like eating a hearty meal, afterward we are no longer motivated to do anything. When we are anticipating rewards, we are motivated to take action. We are driven to take risks in pursuit of them. This process is energizing and often enjoyable. When we are curious and exploring, we often get the best of both worlds:

- The anticipatory excitement of seeking new rewards
- The fulfillment of consuming them

Using brain scans, in study after study scientists found that the striatum was an inferno of activity when people didn't know what was going to happen or if they expected something rewarding. If dopamine responds to the possibility of rewards, it should energize us to seek out these same rewards. And this is exactly what happens. Before people explore, in hopes of being rewarded, the striatum lights up. It doesn't matter if we approach an attractive person at the bar and are told to $%#@ off or if they give us their phone number, the striatum lights up before we make our move.

This is very important—our brains prepare us to take action. When we are in a novel, uncertain situation and notice something pleasurable that we want (think of Charlie and the other kids walking into Willy Wonka's Chocolate Factory for the first time), we can see a surge of dopamine activity in the striatum. The striatum prepares our body to capitalize on these novel opportunities.

Novelty+Open and receptive attitude=new information being integrated into the self and personal growth.

Our body is prepared by directing our attention to focus on the present, mobilizing energy in case we require intense effort in an upcoming challenge, and initiating movements to inch us closer to the source of novelty and any rewards that we can absorb from it. A beautiful sunset or piece of music requires us to focus our attention in its direction and sustain it there; to capitalize on an intriguing comment by someone else, we need to vocalize our thoughts and coordinate our smile and posture to show we're interested.

This is an important discovery: Not all rewards are equal in releasing dopamine from the striatum. When we recognize something as novel, uncertain, or challenging, and we push beyond the boundaries of what we know, dopamine is produced and released at a greater rate. Furthermore, when the novelty is personally meaningful or important to us (picture a five-year-old who can't sit still, pleading to open Christmas gifts), there is a greater cascade of dopamine.

Curiosity Feels Good, Sometimes: Taking Stock of a Few Curiosity Patterns

Being curious is not just a neutral, neural process—our feelings come along for the ride. Sometimes there are joyous, exciting feelings involved with our curious adventures (hiking along an inactive volcano just as the sun is rising), and this means that in addition to an active striatum with a dopamine explosion, you are also going to see a surge of opiates in the brain. Opiates are the neurotransmitters that are most central to that basic feeling of enjoying the rewards life offers.

Curiosity Profile #1: Dopamine with Opiates

Judging by dictionary entries and everyday conversation, curiosity typically refers to a joyful, excited state. We enjoy the uncertain and novel because it piques our interest. We are drawn to the unknown by a sense of wonder and intrigue. Other times, we intentionally seek out novelty, variety, and challenge.

A profile of curiosity and joy is common. Consider a parent listening to a child enthusiastically describe a field trip, a teenager losing herself in dance to an unfamiliar song on the radio, or being immersed in a captivating conversation, book, movie, or game. Most positive experiences contain some degree of curiosity and exploration. It doesn't matter whether it's music, dancing, yoga, sports, leisure reading, movies, hiking, traveling, intimate conversation, good ol' fashioned play, or being engrossed in an activity at home, work, on a plane, on a train, or anywhere.

Emotional patterns of curiosity are visible in our brain circuitry. If we had access to the interior of our brains when we are consumed by joyous explorations, what we would see is an abundance of dopamine and opiate activity.

Curiosity Profile #2: Alone with Dopamine until Much Later

But joy isn't always part of the picture. Other feelings such as tension and frustration are often mixed in with our curiosity. We might agonize over gaps between what we know and don't know. *What's that thing in my eye that is making me blink madly? Of all the cars in the parking lot, why do pigeons choose mine?* We can see the distinctiveness of this emotional and motivational pattern of curiosity in the brain. The dopamine is there and the opiate activity arrives much later in the process (if at all).

When our curiosity is sparked by a need to know something, the pleasure results from getting the answers and feeling a sense of completeness. Only after we fill the gap do we feel better, whether it's feeling calm, less anxious, excited, or some other emotional blend. Just think of those rage-filled moments when you can't get your computer to cooperate. You try to problem-solve, you stew as you wait on the phone for a computer technician, and neither of you can figure it out for several frustrating hours. You are interested in what the hell is going on, but I don't need to tell you that joy is nowhere to be found. Satisfaction arrives, sometimes, when the problem is resolved.

This is a pattern we find in all sorts of situations. Daters are curious about whether the person they went to dinner with feels the same way they do, teachers are curious about what types of rewards and punishments will tame class clowns, and managers are curious about what accounted for last month's wave in productivity. We are still curious, but in the end, we want to get our piece of information and extinguish our curiosity. Dopamine cells fire in response to seeking out novel and uncertain information, even if we are not excited at the time or when our cravings are a bit uncomfortable. In this circumstance, we expect to see a lot of dopamine activity but little or no opiate activity in the brain.

This profile of curiosity with the goal of relieving the tension of not knowing something is probably just as common as that joyous sense of wonder. Besides sharing the common feature of wanting to investigate and explore, they also serve as gateways to making sense of the world. Sometimes it's a small pocket of meaning, and sometimes we discover a profound revision in our approach to seeing the world or living. Of course, if we can't retain these experiences, the end result is nothing more than transient rewards.

The Role of Memory

There is one more aspect of the brain to add into the mix—the hippocampus and its role in memory. Dopamine in the striatum, whether it is actively firing like a machine gun or lying in wait, is central to curiosity and exploratory behavior. If you eliminate it, you become an apathetic, lazy lump. If you increase it, you become fully engaged in the world, searching for and discovering new things.The peak dopamine activity in our striatum occurs before we act on our curiosity, when we are still eagerly deciding whether and how to capture rewards that beckon us. It remains unclear whether dopamine circuits alone account for the learning and growth that occur after we experience and embrace the new. It appears to require help from the hippocampus.

Most positive experiences are ephemeral, so fleeting that they do little more than give us a jolt of goodwill. For a positive experience to last, we need to transfer information into our memory banks. We need to pay attention, take an interest, and ensure that it stands out among the backdrop of other stimuli clamoring for immediate consideration.

Why do we only remember some of what we encounter and explore? Why do we remember some events in exquisitely rich detail and forget others entirely? Part of the answer is whether a portion of the temporal lobe called the hippocampus is energized. It is the main location for the transfer of novel information and experiences into long-term memory, and it is crucial for us to create new memories.

Insights about the hippocampus arrive from years of experiments and surgical procedures. People with damage to the hippocampus have problems recalling specific personal experiences. For us to sustain curious experiences so they can become part of our long-term identities, guiding us in the direction of fulfillment, we need the hippocampus—a vital portion of the limbic, or "emotion," system of the brain.

When the circuitry in the hippocampus is working on all cylinders, we can do more than simply recall a positive memory, we can remember vivid contextual details surrounding the event. When and where did it take place? Who else was around? What were we thinking and feeling at the time? How did things appear? With these rich details, we can savor curious moments and learn from them. Besides being crucial for remembering the backdrop of what happened to us in the past, there is recent work showing that the hippocampus is essential to imagining detailed scenarios of what might happen to us in the future. Seeds of creativity and imagination are sown in this small brain region.

If the hippocampus is activated at the same time you eagerly anticipate rewards, then we don't just feel curious and want to explore. We are better able to hold on to these moments, thus expanding our sense of self. Attempts to find a single neural circuit for curiosity are futile. Dopamine, opiates, and the hippocampus create the brain ensemble that allows us to gravitate from fulfilling moments to a lasting, rich, meaningful existence. Appreciate the importance of getting this ensemble to play together.

The Curiosity Advantage

The greatest advantage of curiosity is that by spending time and energy with the new, increased neurological connections are made possible. Facts and experiences are synthesized into a web, paving the way for greater intelligence and wisdom. We become more efficient when making future decisions. We become better at visualizing the relativity of seemingly disparate ideas, paving the way for greater creativity. It is the neurological equivalent of personal growth. New pathways in the brain are inevitable when you seek out new information and experiences and integrate them into the previously known.

The Curiosity Advantage

4 The Curious Moment
Sparking Intrigue in the New
and Meaning in the Mundane

A fulfilling life begins with interesting, meaningful moments. In 2007, the Princeton economist Alan Krueger, Nobel laureate Daniel Kahneman, and their colleagues published an eye-opening paper called "Are We Having More Fun Yet?" As a result of the massive social progress, economic prosperity, and technological advancements of the past 50 years, an endless array of options has opened up for how to use our free time. In the background was a lingering question: Did these new opportunities allow us to spend more time doing what we cared about most, thus increasing our satisfaction and meaning in life?

For too many of us, too frequently, the answer is a resounding no. Most people spend less than 20% of the day engaged in what could be termed meaningful activitiy (e.g., talking with close friends, spiritual practice and prayer, love-making, playing) or meaningful work.

A Fulfilling Life Is Made of Fulfilling Moments

It doesn't have to be this way. Our lives don't have to be dominated by the "daily grind." Curiosity can be harnessed to transform mundane, unsatisfying tasks in everyday life into something genuinely interesting and enjoyable. These "interest enhancing" strategies allow us to intentionally find and sculpt wonder, intrigue, and play out of the everyday events that confront us. Once our interest is aroused, curiosity plays a second role in ensuring that these moments endure over time.

Without being interested in what we do, it is nearly impossible to be motivated to reach the goals that are most important to us. When we are disinterested, our commitment, effort, creativity, perseverance, performance, and a host of other valuable qualities suffer.

These two simple processes—triggering intrigue and sustaining interest—are at the heart of a fulfilling life.

The Pleasures and Pains of Our Daily Routine

Given the near-universal desire to be happy, or at least happier than we are, we are surprisingly ignorant of what contributes to our happiness quotient. Attempts to improve the quality of our lives must begin with insights into how we feel about the activities that monopolize our time. A few questions come to mind: How much time do we spend doing things that we enjoy the most, giving us a reason to wake up in the morning? How much time do we spend in activities that are unsatisfying? With freedom to do as we please, do we spend it wisely doing things that are truly pleasurable and meaningful? Can we make less desirable activities more pleasurable?

Alan Krueger and Daniel Kahneman opened our eyes to the average American's daily routine. Their approach to studying everyday people was groundbreaking. They recognized that each of us is an unreliable reporter of what actually happens in our lives and particularly, how we felt at the time something happened. We distort our recollections of the past, often remembering the good as better than it was, remembering the bad as being much worse that it was, and telling little white lies so that we can make a good impression on other people. So what these scientists did is ask thousands of people to walk them through a single day in their lives. Study participants relived episodes of their day like a movie, shifting from one scene and location to the next, from when they woke up until they went to sleep.

By capturing what actually happens in people's lives, we can diagnose problems, discover our strengths and how to optimally use them, and make inroads into what to accept and what to change.

What We Allocate Our Time and Energy to Becomes Our Life

With these improved scientific tools, we can take a peek behind the curtains of people's lives. To figure out exactly what Americans do in a typical day, interviewers talked with a whopping 4,000 Americans. Study participants were asked to split up the prior day into 15-minute periods and relive what they did, who they were with, and how they felt. We can be confident in the accuracy of the findings because these 4,000 people were selected to represent every part of the United States, matching census data on age, gender, ethnicity, and so on.

With the small amount of time that we will be on this planet, what percentage of our day is enjoyable, interesting, and meaningful? Do we squander our precious opportunities to be alive in the present?

Discovering what grounds us, what is fulfilling to us, we can look for ways to strengthen these moments—doing it more of the time, making them last longer, or being more mindful so that we can better savor them. It's just as important to closely examine our unfulfilling moments. You might be surprised at how much of your time falls into this realm. With insight, we can ask ourselves if we are ready to invest in some attempts to win back some of these unfulfilled moments. Are there obvious places to squeeze in more well-being at work, in our relationships, and in our free time?

I want to share the intriguing findings that emerged. Most of it isn't pretty. In hopes of making these findings meaningful to you, I want you to envision a pie chart with six slices. Each slice represents a different kind of moment that together, characterizes a day in the life of a typical American. Even if it doesn't fit your life perfectly, I am sure there is much to be gained by comparing yourself to the average Joe or Jane out there.

Let's start with the tastiest, most rewarding slice and work our way around to the most annoying, unwanted slice in what I call "the pie of life."

Highly Enjoyable and Meaningful

Activities such as spiritual practices, parties, exercise, being in nature, listening to music, sex, and having fun (however it is you do it) make up 17.1% of the day.

Enjoyable

Activities such as walking, reading, talking on the phone, and being on the computer make up another 11.5% of the day.

And that's it. Take these top two categories together, and you find that only 28.6% of our day is spent doing enjoyable things!

Mental Valium

Another 22% of our day is spent decompressing in ways that fail to bring us joy or pain, activities such as watching TV, snacking, or just "doing nothing." This time amounts to a black hole of action without emotion, without substance, and without any memorable influence.

Mundane Chores

We spend 14% of our waking hours doing mundane chores such as yard work, washing dishes, and taking out the trash. Many of us dread these events before they even start, and these negative feelings only intensify as we begrudgingly get them done.

The Conveyor Belt of Work and School

The largest chunk of the pie is the 31% of the day that is monopolized by working or going to school. Please keep in mind that this is the *average* for the 4,000 people interviewed. Let's not ignore the various ways in which people view school and work. As I will detail in the next chapter, some of us despise work, doing it merely to make ends meet. For others, work provides an ideal outlet to capitalize on our strengths, providing a meaningful extension of our identity. Arbitrary distinctions between work and play are blurred. Carrying around a small hand-held computer that randomly signaled workingmen and -women throughout the day to report on what they do and feel, a surprising result emerged. The most intense moments of joy and interest occur three times as often at work than at leisure. People with a calling or purpose know that work is not something to dread. Others miss out on these moments and put fulfillment on hold until the workday is over. Thus, how you feel about this slice depends on your attitude toward work or school.

Hurt and Pain

For the vast majority of Americans, nearly 20% percent of every day is spent in unsatisfying activities; this is the most unwanted slice of pie. From commuting to work, fixing broken appliances, and waiting on line at the DMV, on average we spend a fifth of our waking life in pain. This is far too much.

Thoughts on the Pie of Life with an Eye toward Acceptance and Change

Research teams use different questions and arrive at slightly different conclusions about what percentage of our time is unfulfilling. What is beneficial about this particular study on the "pie of life" is that a massive number of people representing the entire United States recounted exactly what happened to them in a single day.

Our relationships with family, friends, and colleagues; our work lives; and our private time are built of moments that are dreaded, easily forgotten, or enjoyable and significant. We possess a limited number of moments. Throwing them away on unwanted and undesirable events is no way to live. We don't have to passively accept our lot. Our mindset, under our control, can be changed. Activities can be modified, even in small and subtle ways. Selecting new, alluring activities and adding them into our daily regimen can be invigorating.

The pie of life offers insight into where to target our efforts. Much of what can be transformed lies within the 31% of our day devoted to work and the 14% committed to mundane chores. Then there is the "mental valium," 22% of the day, when we decompress in front of the TV, snack, or just sit around doing nothing. Don't get me wrong, sometimes recharging our batteries is paramount to anything else. When we get that desperately craved free time, however, we often squander it on habits and routines that might offer a respite from hard work but fewer rewards than we realize. As if all this wasn't enough, inane small talk dominates far too many conversations (even though everyone wishes they could talk about something else). If you fail to properly feed relationships, the fibers of intimacy can easily erode.

Our impressions of how good we feel is often far removed from reality. Human beings possess terrible insight about what brings them happiness and how they are going to feel in the future (see Chapter 1). Collecting and assessing information about what we do in a given day and how it actually benefits us (again, less than a third of a typical day is fulfilling!) is stage one for creating a well-lived life. With an open and receptive attitude, let's probe a bit further and investigate the best and worst events in a typical day.

In this same study of 4,000 people, nothing is scarier than the razor-thin period of time devoted to the five activities judged to be the most enjoyable. Consider this, men and women spend only 25 to 30 minutes per day doing what they love: playing with children; listening to music; going to a concert, movie, or sporting event; being in nature; and partying with other people. Think about it: every day we spend a half an hour doing things that bring us the greatest joy. To put this in perspective, we spend about 40 minutes every day doing the five things we despise most. And if you think it's just one study and these researchers must be mistaken, think again. In other studies (with nearly 1,000 women from Ohio and another 1,000 from France), people's ten most enjoyable activities added up to less than 25% of the day. Clearly, we're not screaming "carpe diem!" when we wake up in the morning.

Consider this a call to arms. We don't need to be slaves to unsatisfying moments. I want to show you the problem, but more important, I want to show you another way to live.

Getting Honest about Labeling Time as Pleasure, Pain, or the Gray Area Between

Freedom—and fulfillment—originates with knowing how we spend our time and, equally important, how exactly it makes us feel. If we remain ignorant of this crucial information, we won't know how to work toward a well-lived life.

An excellent case in point is the experience of parenting. Whether or not we have children, all of us know that parenting is not all hugs and giggles. Then there are the heaps of dirty diapers, children waking in the night, inconvenient doctor appointments, and nonstop caretaking. Yet ask a parent about how much they enjoy being with their kids, and their first response is going to be something akin to "I love being with my kids," "What mom doesn't enjoy it?" I know this from first-hand experience. The more casually you know someone who is a parent, the more emphatic they are that nothing is better than being a parent. I also know this as a researcher. For instance, when researchers asked parents to order a list of activities by how enjoyable they were, childcare ranked in the top two. Only relaxing was deemed more pleasurable. Spending time with friends, talking to friends on the phone, and exercising were all ranked as less enjoyable.

Yet despite what we might say and what these research participants listed, when parents are asked about specific moments in their days, it turns out they enjoy spending time with their kids *substantially less* than they think.

When these same parents were asked to reconstruct a day and consider how they felt during specific episodes with their kids, childcare dropped down to 12th place. The only activities viewed as less enjoyable were commuting to work, housework, and working. Now, this doesn't mean childcare is terrible, it means that with an honest assessment, moments of intense joy switch off with moments of great strain and feelings of being trapped. Taking a closer look at the real picture doesn't diminish the profound meaning that many parents gain from raising children. These findings normalize the pain and discomfort, giving parents the freedom to acknowledge how they truly feel. After having children, most parents experience a major drop in pleasure until the kids leave home in early adulthood. But positive feelings are only a small part of our well-being, and children open up other avenues such as incomparable, lasting meaning and purpose in life.

If you want fulfillment, it's time to stop being diplomatic and confront the truth: Not all moments are created equal. On average, parents find very little pleasure when they are with the kids. Why? Because the lowest lows with our kids, frankly, suck. Think of your own child, niece, or grandson belly flopping onto the supermarket floor, crying and flailing, ensuring the pity and evil-eye scorn of onlookers. But the highest highs with our kids are incredible. Just think about how you feel when Ray Charles's "What I'd Say" plays in the background, and, without the slightest semblance of self-consciousness, your kids swing their arms, squat up and down, turn in circles until they get dizzy, laugh hysterically, and basically look like a bunch of oompa loompas on speed.

What happens when you average these diametrically opposed experiences? Frequent, unpleasant, exhausting, disorienting, sleep-deprived moments interspersed with moments of blissful perfection. Averaging these moments together fails to capture the truth of parenting. Even if our kids easily wear us down, there are strategies we can use to help us feel profoundly more alive and engaged more of the time. Acknowledge that some moments are more of a trial than others, and you are well on your way.

Look at your life and observe what is happening from one moment to the next—the mere act of paying attention may snap you out of your trance [insert image of masked kids on the conveyor belt in Pink Floyd's *The Wall*]. It may inspire you to take a walk around a lake, call a close friend, or have a tickle fest with your kids or current love interest. That is, injecting some passion into your day. Right now, you can start plotting ways to shrink or transform the parts of your day that are most onerous.

Not All Moments Are Created Equal:
The Five Kinds of Moments in Everyday Life

There's an inspiring message in this somewhat depressing news about how most of us spend so much of our time. We can measure and then shift the balance of desirable and undesirable moments in our lives. Before we can add activities that optimize our well-being and remove activities that detract from it, we need to penetrate the motives behind our movements. Namely, what is it that allows us to do what we love and why is it so hard to break free from what we despise? What we do and what we might do can be categorized into five kinds of activities, each with a different impact on our quality of life.

Activities we truly enjoy and naturally want to continue doing

Activities we don't enjoy, that aren't necessary, and take a toll on our lives

Activities we don't enjoy, but for important reasons, are necessary to continue

Activities that allow us to rest and replenish ourselves

Activities we don't engage in that might contribute positively to our lives

Some activities are inherently interesting and enjoyable on their own, and we don't need prodding to continue engaging in them. If you are passionate about achieving a specific goal—large or small—you'll be motivated to achieve it. Your interest, engagement, and curiosity will carry you forward.

Our hobbies, interests, and bursts of play fall into the category of desirable activities. (So does work when it is your calling, or purpose, in life.) During a task, the more curious we are, the more willing we are to exert effort and energy to engage in it. However, it doesn't *feel* like effort because when we are absorbed and motivated, the effort is enjoyable instead of draining and aversive. When we are curious, the critical point at which our energy is depleted is pushed back, and we can persist at tasks for longer periods of time.

The interested person naturally has greater stamina and capability, and interest is a powerful predictor of effectiveness. Managers, supervisors, coaches, teachers, parents, and even relationship partners would be smart to heed this advice: delegate tasks to an interested person.

As for what affects our interest in sticking with something, there are a number of important factors. *Tabula rasa* is a myth as we are not an empty slate. When we take up an activity, we carry beliefs about how interesting it appears to be. These beliefs are going to influence whether we stick with it or not. We also carry beliefs about our ability to handle new, uncertain, or challenging situations. Expertise is less important than what we believe about our skills.

Imagine entering a casino. The baccarat table might appear novel and interesting. It is viewed as the most sophisticated game. Players feel like royalty, as it is often played in a back room behind velvet curtains. Without confidence in your ability to learn how to play the game, your initial interest is sure to be replaced quickly by an unwanted surge of anxiety. For curiosity to reign, situations with high levels of uncertainty require a matching level of self-efficacy (the belief that you can be effective).

Some activities aren't personally interesting or valuable, and continuing to do them can be hazardous to our health. Deciding to quit is often the best possible way to deal with activities and situations we don't enjoy. Quitting reduces stress and conserves our energy for more valuable and engaging outlets.

Consider this radical idea: There is absolutely no reason to continue doing activities we don't feel compelled to do. And if we do feel compelled, we can pay careful attention to "how" and "why" we feel that way in order to recalibrate the ratio of pleasurable to painful moments in our daily lives.

We often allow ourselves to get stuck in unwanted activities because of erroneous beliefs and the motivation to avoid pain (as opposed to seeking rewards). Your favorite band is coming to town and you were planning to buy tickets, but something else that isn't compelling pulls you away. You might be saying to yourself:

If I don't help my brother-in-law clean his garage, my wife is going to be pissed.

If I work this weekend, my boss will know he can count on me, and he'll think I have management potential.

I shouldn't be frivolously spending money on myself.

None of us are immune to avoiding the wrath of a loved one or trying to make a good impression, However, if the toll is too great, if we allow ourselves to continue to deviate from our cherished values, we suffer needlessly.

Some activities are uninteresting and painful, but there are reasons we feel compelled to continue doing them. It would be wonderful if every moment of our lives was exciting and interesting or, alternately, calm and serene. Unfortunately, an average day is filled with mundane and even undesirable tasks that cannot be escaped. Cleaning dishes, pumping gas on a blustery winter day, and calling credit card companies to fix mistakes on last month's bill are necessary aspects of modern life.

Consider child-rearing again. There's nothing interesting or exciting about changing a dirty diaper, especially when the wearer kicks and cries during the ordeal. Yet that diaper has to be removed. Successful parenting is a process, and as much as you want to steer clear of the drudgery of the job, diapers are part of it. Hopefully, if you are a parent, you view parenting as valuable and important. You don't quit after watching a little hand swipe a bowl of soup onto the carpet and walls. Roused at 4 A.M. to soothe a nightmare-induced wail, you endure because you care.

But as the research studies presented earlier show, the activity of parenting is stressful and it would be disingenuous to say it doesn't cause wear and tear on our bodies, other relationships, and that joyous sense of freedom that existed before the kid(s) arrived.

Whether it's childcare, working for a living, spending time with other people, or getting along with our next-door neighbors, there are painful, dull, and mundane moments that are as frequent as the interesting and exciting ones.

Important, valuable activities that are unsatisfying put us in a bind.

We can quit and face real, problematic consequences. For example, we can leave an unsatisfying job and risk not being able to pay the rent next month. We can continue onward as is. However, fatigue, anxiety, hopelessness, and anger are the side effects of feeling deprived day after day. Or, there is another option that is cause for optimism: We can alter the balance of satisfaction by making everyday tasks and activities more meaningful and interesting through the catalyst of curiosity, as we'll see a bit later in this chapter, in the section on "Find the Unfamiliar in the Familiar."

Some activities give us downtime to replenish our curiosity-seeking energy. Going back to the typical day, most of us spend nearly a quarter of every day in mind-numbing downtime. It would be a fatal mistake to argue for the need to be curious and open every moment of the day. Although we feel energized when curious, we need to build our energy supply to tackle the challenges the world offers.

When you are running or lifting weights, muscles eventually fatigue and require rest before they are at maximum strength to begin anew. Contrary to popular belief, exercise doesn't lead to muscle growth. It is when we are relaxing and sleeping that microscopic tissue tears are repaired and muscles expand and strengthen. If you skip resting, overtraining causes muscles to weaken and you end up worse off than if you do no training at all.

Likewise, pseudo-badges of honor for working longer, faster, and harder than anyone else and getting by on a mere two to four hours of sleep are deceiving. Based on his work on sleep deprivation, the psychiatrist David Dinges explained in a *60 Minutes* interview:

There's a cumulative impairment that develops in your ability to think fast, to react quickly, to remember things. And it starts right away . . . a single night of four hours or five hours or even six [hours] of sleep can in most people, begin to show effects in your attention and your memory and the speed with which you think. A second night it gets worse. A third night worse. Each day adds an additional burden or deficit to your cognitive ability.

Distinguishing habit from effectiveness cannot be overlooked. Just because we do something regularly doesn't mean it serves the purpose we think it does. Based on Nielsen media research collected in November 2008 in the United States, the average household has the television on for an incredible eight hours and 18 minutes per day! In the United Kingdom and other parts of the world, it isn't much better, as the average person watches four hours of TV per day. When asked how they feel, it's not pleasure or relaxation; more often than not it's a state of dullness or mild discontent. Whatever the function of television is, it usually doesn't provide the replenishment we think it does. The same might be true for other pieces of your daily repertoire.

When you pick an activity to replenish your energy, it shouldn't feel neutral, it should feel good. Replenishing activities induce positive energetic states, such as excitement, or calmer states, such as serenity and tranquility. What matters is that energy is produced rather than expended.

We need energy to act on our curiosity, and we need curiosity to discover what works to replenish our energy. We can observe, experiment, and modify our relaxing and refueling activities as necessary to discover what works for us. This exercise should be nothing more or less than a personal trial-and-error. Instead of going on hunches and assumptions, for one week use a notebook to record how you feel before and after an activity (interested, energized, relaxed, bored, tired, irritable) on a scale from 0 (didn't feel it) to 10 (very intensely).

Honor your individuality. You might be extremely extraverted (gregarious, assertive) or introverted (reserved, quiet). Some people recharge by being with other people, and some people do it alone. Don't do what you *should* do (because of your gender, age, or career) or what other people do; follow your own sense of wonder and interest. If you take a look at a two-year-old, you don't find limiting categories of what interests them. Aberrant as it seems, some teenagers find cleaning dishes to be relaxing, some bodybuilders unwind with pedicures, some construction workers fuel themselves with a book by Jose Saramago or Milan Kundera, and some homemakers recharge their batteries by riding 10 miles of whip-lashing mountain bike trails. I know because I met some of these characters. Abolish categories, labels, and stereotypes. Artificial barriers only prevent us from knowing ourselves, making choices that fit our interests, and obtaining the required energy nutriments to expand and grow.

Unaccounted-for activities that are not on our radar (they've been rejected or never considered). The prior four categories of moments share one thing in common: We actively take part in them. Unfortunately, there are many other activities that we don't even consider trying in the first place because we assume without any firsthand experience that they will bore us, cause us misery, or be a waste of time. Perhaps we learned this from our parents or friends or by listening to other people we respect or admire.

Consider the "machismo" and "rules of the street" that characterize many ghetto cultures where protecting honor is of utmost priority. Men from these cultures learn from role models that being willing to physically fight others, avoiding acts of weakness such as saying "excuse me" or expressing gratitude, and viewing authority figures as the enemy are the markings of the strong at the top of the social hierarchy. Embracing these views means discounting a lot of activities that might be interesting and enjoyable. We can see similar behavior among people who try really hard to act the part of the professional, the slacker, the jock, or the intellectual, instead of following their own values and interests, which might cut across seemingly impermeable boundaries. We lock ourselves into views of what is "good" or "bad" by making these premature commitments. There are times when we are curious about something, but we hold back and don't even consider acting on it, leaving us with what I call "unsatisfied, residual curiosity." We deny ourselves the opportunity to do the things we are most interested in, and we end up with regrets and a disconnect between what we are doing and what we want to be doing.

The solution for creating the largest area to roam free and find pleasure and meaning in life is to be open and to observe, experiment, play, and revise as needed. The worst that happens is that you get to taste a part of life once, and the best that happens is you find a new and lasting source of interest, excitement, and meaning.

Many of the combat veterans returning from the war in Iraq find themselves having difficulty transitioning to civilian life. They suffer from nightmares, intrusive thoughts about almost being killed and seeing friends get killed, and other impairing anxiety symptoms. I spoke to a veteran at Walter Reed Army Medical Center as he scoffed at his therapists' suggestion that he take part in their three-week yoga program. He said, "It's for sissies, not warriors." Two weeks later, having started the program, he laughs when I bring up our first conversation. What was once completely off his radar, something he despised and ridiculed, became a step out of his deepest pain and despair.

Who knows what each of us is missing out on by our lack of curiosity and exploration. It might not even be a brand-new source of interest; it might be recognizing the unfamiliar in the seemingly familiar. By taking a step back and looking at something as if it is the first time, we see things in people, places, things, and tasks that we might have missed by dismissing them too quickly or prejudging what we expected them to be. We can use our "unborn Buddha mind" (see Chapter 2) to be open and receptive in the present moment.

Awakening Interest: A Missing Piece for the Fulfillment Formula

You might think that it's all well and good to talk about spending more time doing exciting and satisfying things but rather dubious about the likelihood of finding such interest in the drudgery of your own life. That's because there's an important missing link in what we've talked about: triggering interest. If we look at five of the six kinds of moments that make up a given day, only one has to do with things we are already interested in or curious about. Before embarking on the quest to create lasting passions, we need to spark interest in the first place, and that spark can be hard to find and easy to extinguish. As the famous education reformer and psychologist John Dewey wrote in 1913, "You need to catch interest before you can hold on to it."

The major hurdles are how to persist at tasks that are not inherently engaging but necessary and how to encounter more situations that have the potential to engage us. Instead of resigning ourselves to lives in which 75% of the day is less than enjoyable or at best, neutral, we can find ways to enliven and transform these moments.

Changing Perspective: Manipulating Situations and Tasks

Starting with Mood

For activities that seem to be uninteresting initially, being in a good mood is often the first spark for catching interest. You want a rich, exciting moment that is arousing, one that captures your attention, and opens your mind to what is being offered as opposed to what is expected. From the obligatory (can't get to work without commuting), to the important (those dishes can't sit in the sink forever), to the inescapable (the boss, who makes you uncomfortable, sits down at your lunch table only seconds after you arrive), enhancing your mood can be the flint that ignites a flicker of passion. It doesn't have to be complicated. You might call a favorite person to get inspired before you begin.

If you are in a good mood, you will often use this information and figure that if I feel good, then I must be doing something interesting and fun. (Note that we're often wrong because sometimes our mood is affected by the weather, hormones, or something unrelated to what we're actually doing.)

A good mood is often contagious. If your boss strolls into the office and offers free sushi to you and your coworkers, those positive feelings (assuming you like sushi) can carry over into your work. Whether you know it or not, you often return this act of goodwill with a pleasant attitude that translates into persistent hard work. Your boss probably knew that before he bought that first unagi roll.

When we feel good, we are more interested in whatever grabs our attention, and we see the world around us differently, discovering that much of what surrounds us is more captivating than we thought. We are more likely to view difficult situations as challenges instead of threats to our safety or ego, and generally we enjoy ourselves a little more. Let me explain what I mean by this distinction:

- Situations are viewed as *threatening* when we believe the demands of a situation outpace our available resources (abilities, social support, etc.), and this is intimidating and scary.
- Situations are viewed as *challenges* when we believe the resources needed to cope effectively are at our disposal.

When we are in a good mood, stress is transformed into interest. We are more willing to funnel energy and effort into activities. We essentially become the ideal worker, student, parent, friend, and athlete that everyone wants on their team.

Engaging Our Strengths

We can actually focus on *increasing* the challenge of situations to engage our strengths. When we use our strengths, we feel energized and alive. We experience a sense of flow, interest, and immersion. We aren't self-conscious; instead, we are focused on the task at hand and perform at our best.

I had an opportunity to work with a client in therapy named Fred, who intentionally created space for his strengths to surface in his work.

Fred is a janitor at Grand Central Station in New York City. For an unskilled worker, he earns a decent salary, but as one of the busiest areas in the city, Grand Central isn't a pleasant place to work as a janitor. For years, he was embarrassed by the repetitive, dead-end work that he says "defines him." Eight-hour shifts of cleaning the station to perfection undone by careless, garbage-tossing commuters and the bodily fluids of homeless people. His actual work is far less interesting than how he ultimately deals with the repetition and frustration.

We talked about his strengths, and of the many that were apparent, one stood out: He could create games for any household or street gathering and by doing so, raise people's spirits. It never occurred to him that he could use the same mindset at work. He transformed the acts of mopping floors, emptying trash barrels, and plunging toilets into a sport. At the start of each task, he timed his performance and kept a tally of his best times in a small, heavily creased, spiral notebook. If there were people in his way or a huge mess to be cleaned up, he gave himself a handicap to account for the increased difficulty level. These calculations and competitions made his tasks far more interesting.

When he explained the system to his coworkers, they laughed. But Fred's personal challenge had set something new in motion. A few days later, some of the others started keeping their own tallies. They teased each other as they wrestled for the position of top dog, trading stories about obnoxious people and grotesque "handicaps" (you can imagine these janitors strolling through Dante's third circle of hell—"the gluttonous" dump of waste and refuse). These playful interactions magnetically drew people together who previously worked alone, and they provided vitality and more initiative for completing assignments. Leaving work, they weren't the depleted, walking dead that family and friends were accustomed to seeing; they now had enough stamina to engage in pleasant activities before work began anew. By generating curiosity in his actions and coworkers, Fred paved the way to draw rewards from a seemingly lifeless environment.

(Re)Connecting with Play and Playfulness

We can add actual play and playfulness to almost any task. This aspect of triggering interest was captured wonderfully in a story on National Public Radio about an assembly-line worker in a potato chip factory whose job was to ensure that chips rolling down the conveyer belt were uniform and aesthetically pleasing before being bagged. Mishaps were to be removed (something a machine can't do as well as a person). He didn't take the job because of the excitement. He took this dreary job because money was tight and he needed the work. What intrigued me was that he developed a quirky game that made the job more interesting. He searched for potato chips resembling famous people and kept a collection (imagine silhouettes of Elvis, Charles Manson, Marilyn Monroe, and the flaming guitar of Jimi Hendrix in the guise of tortilla chips). Constantly scanning odd and bizarre shapes for celebrity resemblances, the day moved quickly and he was incredibly efficient at catching mishaps.

At Trader Joe's, cashiers are trained to ask open-ended questions and to follow up on customers' replies. The idea is that when customers and workers are interested, everyone gains. It also has the additional benefit of creating an infectious good mood. Instead of the standard "how are you?" customers are baited with any of several openings. Anything exciting going on in your life? Did you get to see Michael Phelps swim last night? What'd you think? So what are your plans for this weekend? Sometimes I try to make them squirm by giving one- or two-word answers, but they're persistent and usually draw me into an enjoyable, personal exchange. It's one of the reasons I keep going back—along with thousands of other happy shoppers—and they know it.

These are just examples. The important point is that we work harder and longer and are more committed and responsible when we are in a good mood. As long as the task gets done, be open to innovative ideas for how it can be accomplished.

The opposite happens when we are uninterested and disengaged. If you find yourself in an unsatisfying situation that you have to get done (perhaps those mundane chores at home or repetitive tasks at work), play isn't just going to feel good, it's going to make you better at whatever you are doing.

Choose the Right Guides and Partners

The right guides and partners can make all the difference in finding the ideal entry point. They can trigger our interest and encourage us to experiment with new activities, people, and domains. Their positive moods are contagious.

Whether a guide is a friend, family member, or coworker or an unfamiliar instructor or group, we do best when a guide supports us without judging, evaluating, or criticizing. Entering into the new often brings a mixture of the positive and negative. People who understand uncertainty and the complex blend of emotions that naturally arise in new situations are good guides.

The right guides and partners set the stage for us to experience new situations as challenging, not threatening. People are more engaged, willing to explore, creative and authentic when they feel safe and secure. If the world is likened to a game of tag, they provide a secure base that we can return to for pep talks or comfort or to share our adventures. The stage is set for us to act on our curiosity and employ our strengths.

I also suggest that you seek out and surround yourself with people who resemble the type of person you aspire to be and who have found success and fulfillment along a path similar to the one you're on now. The more you learn about the serpentine paths they took and the setbacks they faced, the more likely their stories will inspire confidence in your own coping skills. With greater confidence, you will naturally be more interested, engaged, and willing to seek out the new and stretch beyond your established comfort zone.

We can evaluate everyone we interact with in terms of how much they support our explorations and experiments that led us to who we are and the future to which we aspire. The right guides and partners accept our unique qualities and our deviations from the norms of social groups and the larger community. They give us the acceptance we need to carve out our niche. To harness our curious spirits, we need to be honest not only about ourselves and how we invest our time and energy but also about the people who surround us. You can ask yourself straightforward questions about the people around you:

- Does your romantic partner show an interest in your past romances, which set the foundation for the two of you to hit it off, or is he or she dismissive and jealous?
- Does your current circle of friends respect tales of when you were younger, perhaps more impulsive and reckless, or are they critical and judgmental?

- Do long-standing family members and friends respect how your personality and interests change over time or do they desperately cling to their initial impressions of you?

Be honest in your answers or you will only reinforce unneeded barriers to growth. Bands like Metallica and The Clash had millions of doting, frenetic fans after their first records were released. As these bands matured, many fans revolted. They wanted the same band and the same kinds of songs for eternity. The "true" fans respected these natural changes. After all, why stay in a band if you're only going to keep doing the same thing? They accepted that if we evolve, other people evolve, too. Change is an inevitable part of life.

Surround yourself with guides who can handle it instead of struggle with it. With the right guides, we can discover and explore much more terrain than we could on our own. It is a sign of strength, not weakness, to have this base of security.

Find the Unfamiliar in the Familiar

For some parents, it's fascinating to watch their toddlers find frays in the carpet and paint specks on the wall as if they are on a hunt to find all the little things you never notice. For others, it's just one of many things their children do and these tiny epiphanies go unremarked until someone else points it out. What is the difference? One distinction is each parent's willingness to notice the new in the seemingly familiar and obvious aspects of everyday life.

I say "seemingly" because recurring situations may look identical on the surface, but in reality any event involving people has some degree of novelty each time it happens. We can be open to this novelty or we can classify it away as just another moment when my child plays in the kitchen, my romantic partner sings in the car, or my boss compliments me on my work.

One way to become more curious is to intentionally circumvent expectations, categories, and labels about "seemingly" familiar activities and events. It's far too easy to prejudge an activity because we think we have seen it before, or avoid an activity entirely because we expect it to be unpleasant, like the man who initially rejected yoga to help with his post-traumatic stress. Removing his resistance, he discovered a path to well-being.

When I ask my students and clients what kind of music they like to listen to, I often get the same response—"Everything but country and rap." (If you happen to be one of the millions who like country or rap, then insert "heavy metal," "Celtic music," "boy bands," "anything resembling Michael Bolton," or whatever genre is the equivalent of a dental drill boring through your molars.) When I ask people which sections of Barnes & Noble they avoid, people often say "poetry, philosophy, or any section where I am stuck re-reading the same page over and over again." When I learn that people don't like sports and ask why, they often tell me, "I have no idea of what's going on and could care less."

Why do many of us have strong dislikes before we even give people, places, and activities a fair chance? Failing to explore, we set up barriers that reduce the size of our universe by a smidgen. We pass up potentially valuable opportunities. The goal of discovering the unfamiliar in the familiar is to suspend judgments and attend to how things are, not how you expect them to be. Research has shown that a small shift in attitude can energize us and change how we evaluate activities we previously viewed as silly, stupid, or unsatisfying.

In an experiment people were asked to do something they reported disliking and they were told to pay attention to three novel features when they did it. This small exercise actually changed the way they viewed the activity. An 18-year-old male bodybuilder who scoffed at crocheting ("it's for grannies"), spent 90 minutes practicing his stitches and discovered that it could be tiring (he overestimated the physical conditioning of his fingers), that it could be meditative ("time flew by"), and that the stitches could be tight enough to create flip-flops (which is what he worked on).

More interesting, when subjects were contacted weeks later, people who were asked to search for the novel and unfamiliar in this laboratory task were more likely to have done it again on their own without being asked or prompted. A window of opportunity opened for a number of participants that not only didn't exist before but was artificially closed off by their preconceived ideas.

There are three steps to applying the lessons of this experiment to our regular lives:

1. Very simply, choose something that is unappealing to you. It could be listening to speed metal or free jazz, going to a wine-tasting event, or reading Victorian poetry.
2. Take part in the activity, but instead of doing it as is, search for any three novel or unique things about it.
3. Write down or talk about what you discovered with someone else. Like the research participants, you'll find you carry this open-minded interest with you in subsequent days and weeks.

This same little experiment can be applied to any activity in your life. Consider the list of low-interest but necessary activities in your typical day. If you are a parent, it might be cooking meals for ungrateful two-year-olds or going over nightly homework with your nine-year-old. At work, it might be mindless data entry or administrative tasks. Living with your romantic partner, it might be figuring out a way to cut expenses from that dreaded monthly budget or planning visits with each other's least favorite relatives.

One of my therapy clients, who suffers from depression, used this strategy to motivate herself to do one of her most despised activities, the laundry. Since she was depressed, she hadn't been motivated to do much more than look at the intimidating pile of dirty clothes that continued to grow, reminding her of how little she achieved during waking hours. I asked her to try washing her clothes just once until we met again. Just to build some momentum. I gave her one other instruction: intentionally search for novelty during the process. When the clothes were dried, she brought the filled laundry basket to her room. Breaking her usual routine, she didn't turn on the TV. With perfectionism, she decided all her shirts needed to be completely flat before she folded them. She noticed that her hands got really dry. She also noticed that her nose started to itch, and every time she rubbed it, it felt worse, like the skin of her nasal cavities was tightening. Then she started to think about where she got each piece of clothing and realized that much of her clothing is made of tough material, contrasting with her quiet and gentle demeanor. It made her wonder about how little she might know about her motives for making purchases. She summed it up best, "I expected the task to take a long time or at least seem like it was taking a long time, but it didn't. It made me appreciate how much time and effort goes into clothing, not just in doing the laundry, but in going out, browsing stores, figuring out what represents me best, and then taking a moment to remember all of the stories and moments that go along with every shirt and pair of pants."

Instead of entering with a closed mind and chalking up events as wasted time before they happen, we can create a little breathing room. If the activity is not part of your ordinary routine, ask yourself what can you find that is novel and unique about this particular activity before stepping in? What can you find once you are immersed? If the activity is part of your regular routine, what is different about it this time, right now, compared to the past?

It's a small quest that prevents us from putting on blinders and falling into ruts. Before I wrote this, I changed the diaper of Raven, one of my twin 20-month-old girls. With twin babies, you change about 10 diapers a day, and toddlers are not the most compliant creatures. Raven usually flips like a flapjack, kicking for dear life all the while. When I search the situation for novelty, I find that she tries to keep her little feet on me at all times (even when she flips onto her belly) and avoids eye contact with me. By gently rubbing her belly, I can usually buy a few seconds of stillness. I often forget to look for the unfamiliar in this regular routine, but when I do look for it, something special happens. Some pleasure enters the atmosphere, and I get a moment to reflect and feel close to my little one. Instead of losing a moment, I gain one.

Amplify Interest by Changing the Way We Remember Events: The "Peak-End" Rule

Our memory is a hall of distortions. Some parts are amplified, others are minimized, ignored, or forgotten. Reality ends up becoming what we make of the past.

We remember our experiences through the prism of expectations. What did we expect? Were those expectations met? This is important because if we remember an event as soothing or exciting, we are more apt to repeat it in the future. If we remember an event as annoying or repulsive, the doors could potentially close forever. This is a lesser-known fact about how the timing of our emotions affects our memory of events.

Few things will get you less excited than a colonoscopy. Imagine two patients, Sylvie and Adele, who both undergo a 30-minute colonoscopy. However, for Adele, after the painful suction process is over, the tube is left in for an extra 5 minutes—it feels weird but is far, far less painful than the first 30 minutues. Who do you think has a better experience?

You would think it must be Sylvie. After all, the sooner you get to put your clothes back on, the better. You might be surprised to know that Adele had a better experience. The reason why Adele fared better can help us all in improving our moments.

The most intense emotional moments of an event (peak pain or pleasure) and the way an event ends (painfully or pleasurably) have a profound impact on how we remember it. Even though Sylvie had a *briefer* procedure, it was the last moments that lingered. For Sylvie it was pain, but for Adele, it was an unusual sensation. We know that endings affect how we recall events from the infamous colonoscopy study conducted by the economist Daniel Kahneman. Half of 682 patients undergoing a colonoscopy had the tube left in by the doctors for an extra few minutes after the procedure was over. Although leaving the tube in for the extra time was less painful than the actual procedure, patients who underwent the longer procedure reported a better overall experience immediately afterward and when asked about it an hour later. Now, who do you think is more likely to return when a doctor suggests they get probed again? There are real consequences to ending on a happy or unhappy note.

Another study of colonoscopies confirmed this finding. The patients in the sample underwent colonoscopies that ranged from 4 to 69 minutes. We would assume that the poor creatures with procedures lasting a full hour would remember the event as being worse than those lucky enough to get out in less than 10 minutes. Yet, amazingly, there was absolutely no relationship between the length of the procedure and the remembered pain and unpleasantness (a correlation of .03!).

Our retrospective evaluations of the events in our lives are best predicted by the intensity of our peak emotions (highest highs and lowest lows) as well as the intensity of our feelings at the very last moment. What this means is that we can change our interest and motivation in an activity by taking advantage of this so-called peak-end rule.

I met people who took advantage of this strategy without knowing it. I asked about how they deal with tasks that fail to stimulate them but they feel compelled to do:

- Shane is trying to lose weight by going to the gym every day, and he ends each workout with a 20-minute soak in the hot tub.

- Because of Type 1 diabetes, Jasmine has to take insulin shots twice a day. Carrying around a pair of dice, she gives herself one roll after every injection. If she rolls a 12, she buys herself a new song for her iPod (only $1). It's clear from her proud retelling that she enjoys playing the game more than winning.

- Mike manages a natural-food store with three cashiers at any given time. He circulates the store every hour and asks customers at that moment whether their cashier should get the honor of picking the next song to be played on the store loudspeaker. The customers get into it and extoll the virtues of their cashier, his employees get into it and are friendly with customers, and the store has an intimate feel that is missing from larger chains. Mike enjoys watching his employees playfully argue about each other's tastes and the musical selections for the day before they head home. Needless to say, morale is high.

Shane and Jasmine show that we can intentionally bring creativity to our actions, and as Mike shows, we can trigger interest and excitement in others. With a little effort we can consciously/proactively create intense highs and happy endings even when we are doing things that on the surface are annoying, painful, or monotonous—and these positive "peakend" experiences can profoundly affect many aspects of our lives.

When You Are Leading and Motivating Others

When I teach my undergraduate and graduate students, I prowl for examples that intrigue them—sex, dating, drugs, violence, movies, music, and food. These examples and situations are effective because they wake students from their slumber. There's nothing like a brief burst of profanity or a lewd reference to bring a student to attention.

I am very mindful of the mood of students and audiences when I expose them to new material—are they apathetic, bored, anxious, or interested? Learning environments that include playful group assignments, surprises, and personally meaningful challenges can trigger that initial surge of interest. Of course, it shouldn't be so challenging that it causes anxiety or so underwhelming that it causes boredom. By allowing people to feel able and competent, they will be less focused on trying to avoid looking like an idiot and failing and, in turn, they can be more deeply attentive to the task at hand. These are ideal conditions for a sense of wonder and intrigue to emerge.

You might also think work and play should be separated by impenetrable boundaries and that there is no room for games, socializing, and food. If this is the case, you need to carefully consider the scientific research that shows that when people enjoy themselves, they are more creative, efficient, productive, cooperative, and are better autonomous decision-makers. What on that list would you not want from your captive audience of children, workers, students, athletes, patients, or anyone else?

Finally, you need to remember the goal. If you want people to be interested, committed, and willing to devote effort to learning, mastering, and using these skills for the long haul, then you can't avoid the initial step of stimulating excitement.

It can be especially challenging to engage people who are marginally interested, disinterested, or unmotivated. When people are doing tasks under your guidance, it helps to be flexible about the ways they can behave and express themselves.

Rigid rules prevent people from finding their own individualized strategies to enhance their interest and only serve to alienate and reduce their performance and commitment. Researchers have shown that when people feel they are being observed and controlled, they experience greater negative feelings and perform worse.

The same goes for intense time pressure, which only disrupts interest because people become narrowly focused on completing tasks and not failing. Loosening the grips of surveillance and oversight and being flexible increase the likelihood of getting elevated interest and dedication from those around you.

An Important Caveat: Being Attuned to the Differences between the Uninterested and the Interested Person

Not everyone enters a task or topic at the same place. Some people start off with a strong initial interest. What happens when you use the same exciting, stimulating techniques with those who are less interested? Sure, many will be transformed and end up becoming attentive, engaged, and interested, but those who start off with a spark of interest actually lose some of their excitement and luster.

This makes perfect sense. Let's say that I absolutely love playing Frisbee but no one is interested in playing a game of Frisbee Golf except my brother, who happens to be a world-class player. To recruit other players, my brother shows us some tricks. He spends 45 minutes showing us how to skip the disc off the ground and up into the target or an overhead hammer throw where it travels flat side down and lands at the target right side up (very weird). Throughout the game, he periodically stops the action mid-game so we can get more pointers. These tutorials might draw the others in, but the tedious down time might dampen my own interest, preventing me from getting into a rhythm.

The same thing happens when you try to teach people about home repairs or computers, or when you try to get anyone involved in a task in which they already possess a strong interest. The bells and whistles can be bothersome and distracting. They can feel frustrated and demoralized that you need to resort to tricks to intrigue them.

So it's important to be mindful of where people are starting from with their interest and background knowledge. Overestimate their interest and you might forget that they need a little extra incentive to be motivated or they will be at risk for dropping out of the picture. Underestimate their interest and you might rob them of pleasures they are already prepared to experience. If you have a mix of interested and uninterested people, one solution is to bring the interested people into the forefront as fellow instructors to show off their talents and potential.

A Brief Primer on Practicing Curiosity

Don't treat your goal of becoming more curious as something frivolous that you can set aside because there are more pressing matters in your life. Health and family trumps nearly everything. After that, few things are as important as your well-being. Approaching life as a curious explorer is going to help you work toward greater well-being. Schedule time for novelty just as you would for work, medical appointments, or anything else that you would never miss unless you had a legitimate excuse. To be more curious, you need to devote time to change.

One of the tried-and-true principles of body building is relevant to becoming more curious. It's called muscle confusion. If you want to transform your body—whether by gaining more muscle, losing fat, or becoming stronger or more agile—you need to consistently alter your routine. Our minds and bodies are designed to conserve energy and do the least work possible. If you go to the gym every day and spend 30 minutes on the treadmill at the same speed, do the same three exercises to build your biceps using the same weight, and do everything in the same order, you can ensure that you will plateau soon. Your body will learn what it has to do, and as soon as it does, it goes on autopilot and although you are still running, jumping, pulling, and pushing, your body stops growing. The key to maintaining a healthy mind and body is to constantly change the variables in your workout. Change the types of exercises you do, the order in which you complete them, the angles in which you hold your body, the number of sets and repetitions you do, the length of rest periods between sets, and so on. Trick your body so it will have to adapt to the challenge. New muscles will be recruited to complete the routines and your body will be forced to grow.

The same goes for becoming more curious. Little things make a difference in changing the way you approach life. It is easy to stick with structure and order because routines make us feel safe and secure in an uncertain world. But we can open our eyes to the fact that novelty and enticing things that can grab our attention are everywhere. We can change our habits, change the way we act, and change the way we see the world anytime, anyplace. Appreciate and search for more than what you already know, already assume, and already expect to happen. I am talking about a mindset of expecting there to be things you don't know and realizing that this does not mean you are vulnerable or unintelligent because you can't predict what is going to happen. Rather, it means there are opportunities for learning, discovering, and growing.

Another simple strategy is to remember to avoid early committal to ideas and judgments. Hold on to the mantra that there is always another way at looking at things, and there is probably something you are paying too much attention to and something interesting that you are ignoring or missing. Keep this in mind every time you make judgments about yourself, other people, tasks, the world, the past, the present, and the future. Wouldn't it be wonderful if this approach were part of the training for doctors making diagnoses, researchers trying to make sense of the world, journalists reporting news stories, businesspeople working on projects, and politicians making decisions? When we are open to new possibilities, we find them. Be open and skeptical of everything.

Practicing curiosity can be difficult. It can't be done once a week, like a psychotherapy session. It has to be done regularly, and sometimes when we are just getting started in altering our approach to living, it has to be all encompassing. We might use the following guidelines:

- When waking, what am I seeing that I overlooked before?
- When talking, I am going to remain open to whatever transpires without categorizing, judging, or reacting, I will let novelty unfold, resisting the temptation to control the flow.
- When walking outside the house, I will gently guide my attention so I can be intrigued by my every bodily movement and whatever sights, sounds, and smells are within my range.
- I will assume or presume nothing except that novelty exists everywhere.

With this mindset, every single gesture is guided by openness and curiosity.

Try it for just five minutes every day. After a week, add a little more time to your training. Ensure that you practice in different circumstances. One day try it at dinner with your family or friends. One day try it at a meeting at work or in class at school. Try it while exercising or while taking a walk, cooking, eating, sitting on your porch, and so on. This approach will ensure you remain flexible, with curiosity unbounded by any particular circumstance.

Stick with this practice and you will be taking advantage of neural plasticity, as experiences change your personality for the better. From triggering interest and pockets of meaning, we move on to the creation of lasting interests and passions in the pursuit of sustainable well-being.

5 Creating Lasting Interests and Passions

Moments that are overflowing with interest and excitement make us feel alive and hungry for more. Yet, most of our curious moments are short-lived. They may be happy experiences that lift our moods momentarily. Or they may be experiences that are simply focused on dispelling uncertainty and finding an answer. Pursuing a good time is nothing to feel guilty about, but strings of unconnected pleasurable moments, are not the secret to a fulfilling life.

After being exposed to an interesting situation, we can catalog it with other positive memories and move on. Or, we can try to hold on to that captivated attention and extend it. This is a tricky proposition. We are talking about creating relatively enduring preferences, interests, and passions for topics, things, people, and places. We are seeking renewable, lasting sources of meaning that can regularly feed our curious appetite. Strengthening and deepening interests is typically self-generated. What I mean is that we might find ourselves questioning, exploring, and creating links to what we find to be important in our lives. This process of drawing connections is started by those initial, lingering positive feelings you get from an interesting moment that you don't want to end. There are strategies that increase the likelihood that curious explorations set the foundation for enduring interest. Enduring interest is largely shaped by the personal meaning and value we find in an activity.

A Taste of Passionate Pursuit

I am reminded of my first intense passion of adulthood. As a gawky, un-coordinated teenager, I could never find a sport that interested me and that I could excel at. These two issues—being good at something and be-ing interested—were hopelessly tangled. Football really interested me. Running at full speed, propelling your body at full strength into the lone person carrying the ball without reservation sounded cathartic. The only problem was that the dream sequence looked different on the field. When a 150-pounder smacks into a 300-pound slab of meat, it shreds vital muscles and tendons. Yet I didn't even have the "pleasure" of hitting someone. In my very first game, I tried to tackle a wide receiver running up the sideline. I dove after him, missed, and smacked shoulder-first into the ground. Within minutes, I realized that I couldn't move my right arm. I spent the next several weeks in a body cast, wrapped in plaster from waist to shoulder with a forearm cemented across my chest.

During a long stint of physical therapy, I remember watching the 1988 Olympics, particularly the shot put. Once upon a time—776 B.C. in Ancient Olympia, to be exact—it was the most popular athletic event in the world. Nowadays, the shot put is typically viewed as a silly thing hu-man beings do to entertain themselves, and the number of spectators it draws rarely exceeds 20 people. For me, it was perfection. One person inside a 7-foot circle with one objective: throw a 12-pound metal ball as far as possible. No matter how far that ball went, no matter whether you won or lost, only one person was accountable. It was athletic purity. It was the perfect test of strength, speed, balance, and discipline.

I became instantly enthralled, and spoke to my twin brother, who was one of the captains of our high school track team, about how to join and whether there were any good shot putters on the team. It turned out that we didn't have a single shot putter, none of the coaches had experience training the shot put, and our high school had been forfeiting this event for years. I spoke to the coach, and my passionate pursuit began.

I started vigorously training my body, lifting weights, drinking wheat grass and protein shakes, and armoring my small frame with muscle. On the clearance rack of an athletic store that was going out of business, I found a neon-green shot put and a polyethylene rope carrier with handles. I suspect this shot put had been sitting on the shelves since the store opened.

Without a coach, I had to figure out how to throw the damn thing on my own. The first thing I did was go through catalogs to try to find the right book. I stumbled upon a small pamphlet on the biomechanics of the shot put with black-and-white frame-by-frame still shots of Austrian Olympic champions. Despite the language barrier (it was in German), I pored over those pages and carried it whenever I practiced. Every day I went to school with my books in one hand, my pamphlet of shot-put biomechanics resting atop, and my neon shot put in the other hand. People referred to me as "the boy with the metal ball." After school, I spent hours throwing that shot put over an unused soccer goal post. I experimented with different ways of holding it against my neck, different ways of gliding my feet across the 7-foot circle I drew in the dirt, and different ways of building momentum, whipping my left leg around in the air and back down to the ground, twisting my torso to garner the fastest speed and quickest release as I tried to find that perfect angle for it to soar through the air.

In the interest of avoiding revisionist history, let me disclose that when I first became a shot putter, I sucked. I sucked badly. But I went back to the soccer field and modified my form, videotaped my movements, and constantly improved my mechanics.

A few things sustained this interest. I began to get better rather quickly, placing and winning in events, and in my second year, I was invited to the state championships. I lost, but I was invited. My exploits were covered in the local newspapers, and I received a lot of praise from the coaches and students who had wondered what I was doing with my spherical metallic pet. I felt confident; I felt a sense of mastery. I set my goals higher for the next season. This passion, this intense interest, bled into other areas of my life. With my new discipline at working so hard for the sake of mastering something, I became a better student. I became more comfortable trying other new things and spending time with new people beyond my few close friends. Shot put became the first of a string of intense passions that fed my strengths and pushed me in the direction of cherished values.

The Intersection of Interest and Meaning

To understand what intervenes between triggering and enduring interest, we can turn to a study of more than 1,000 college students. Before the students began an introductory psychology course, they were asked about their general interest in the subject. Subsequently, these same students were asked at regular intervals about their grades and what they thought about the field of psychology. These check-ins continued for two years and the students' academic records were reviewed to see which courses they took and how well they performed in college.

The scientists leading the study found that students with an initial strong interest in psychology were inspired during the semester, enjoyed coming to class, and because they were interested and attentive, were more likely to apply the knowledge to their own lives. With a stronger personal connection to the material, the more-interested students received better grades, enrolled in more psychology courses, were more likely to major in psychology, and had higher grades in these courses.

One reason that interested students performed so well is that their goal was to master as much about psychology as possible. In stark contrast, less-interested students tried to get by with as little effort as possible to avoid bad grades. Different approaches to that very first psychology course led to dramatically different college careers.

The same patterns are found outside the classroom. In another study, researchers observed nearly 200 athletes at a summer football camp. Athletes entering camp with the greatest interest in football found the greatest enjoyment and value in their daily training. When you are interested in something, the little things matter—practices, scrimmages, and waking up in the middle of the night to run or lift weights. This open, receptive mindset is visible and attractive to others. Athletes rated as the most talented and desirable by coaches were not simply the most physically gifted, they were the ones who were the most interested when camp started.

Whether it is school, work, sports, parenting, household chores, or doing taxes, we are going to be interested and motivated if we can connect our immediate actions to larger, personal values. When we can't, disengagement, burnout, stress, and poor performance are much more likely to occur.

On the Basketball Court: When Meaning-Making Fuels Passion

The athlete Bo Kimble excelled by bringing values into the forefront when faced with a heart-breaking personal loss.

In the late 1980s, Bo Kimble and his best friend, Hank Gathers, held the basketball world in thrall. They went to high school together in a rough neighborhood in Philadelphia and took their school to the prestigious public league championship in 1985. From there, they received full scholarships to attend the University of South Carolina. When the coaches who recruited them were fired, they fled together to a little-known college basketball team at Loyola Marymount University in Los Angeles. While the average college basketball team was lucky to score 100 points, Kimble and Gathers shocked the sports world by leading their team to score 157, 150, and 148 points in three consecutive games over four days.

In 1989 Hank Gathers became the second player in history to lead the nation in both scoring and rebounds with Bo Kimble not far behind him on the list. Then in 1990, in the middle of a game—not just any game, the semifinal of the West Coast Conference championship tournament— right after making a thunderous dunk that drove more than 3,000 fans into a frenzy, Hank Gathers collapsed and died on the court.

When Bo Kimble returned to the basketball court, he led his team through the final tournament of the season with the top 64 teams competing. He brought little Loyola Marymount past some of the best teams in the country to the final 8. He graced the cover of *Sports Illustrated* magazine.

What allowed Bo to focus, what gave him purpose, was to honor his best friend by living out their shared dream. Even though Hank Gathers was one of the best basketball players, he struggled to shoot free throws. At one point, he practiced using only his left hand to get it up to speed with his dominant right hand. After Gathers died, Bo shot the first free throw of each game left-handed in honor of Gathers. He made the first three shots he attempted, which is no mean feat. And just like Gathers the year before, he led the nation in scoring. When he was drafted into the NBA with a multimillion-dollar contract, he continued to honor Gathers by making his first free throw of each game with only his left-hand.

I'm sure Bo's coaches and teammates had mixed feelings about those left-handed shots as they were far less likely to sink, costing the team points, and perhaps led to a few team losses. But Bo's coaches would have been foolish to ask him to stop, and they didn't. By connecting his work to profound values about friendship, his interest, and effort intensified. Besides his own accomplishments, Bo inspired teammates, other teams, and a younger generation to emulate his game and character.

There is much to learn from Bo and his open-minded coaches and teammates. We, too, can learn to connect what we do to our values. If we are trying to motivate others, we can help them do the same.

Values as a Catalyst and Guide to Meaning

Our values influence why we do the things we do, and the rewards of moving in the direction of our values motivate us to persist despite challenges.

Identifying our values is critical to creating a fulfilling life and can begin with a simple questionnaire. It is based on decades of research by therapists helping clients move from thinking about changing to taking the next step of doing something about it.

Some of the values listed below might be irrelevant in your life and some might be the guiding forces that drive your behavior and decisions in nearly everything you do. Layers of interest and meaning can be added to any activity when we can connect our behavior to cherished values. Using the list below:

1. Identify the ten values of greatest importance to you.
2. Circle them, write them on a piece of paper, tuck the list into your desk drawer, or tattoo them on your calves. It doesn't matter. What matters is that you carefully go through each of the values until you find the ones that resonate. Meditate on these values regularly to build a foundation to guide your actions and to heighten the meaning and satisfaction found in seemingly minor activities.

*   **Acceptance**—To be accepted as I am
*   **Achievement**—To set goals and make important accomplishments
*   **Accuracy**—To be accurate in my opinions and beliefs
*   **Attractiveness**—To be physically attractive
*   **Authority**—To be in charge and lead, command, and be responsible for others
*   **Autonomy**—To be independent and in control of my thoughts and actions as opposed to being controlled by outside influences
*   **Caring**—To take care of others and be kind and generous
*   **Challenge**—To take on difficult and demanding tasks and problems
*   **Commitment**—To make enduring, meaningful commitments
*   **Conformity**—To respect rules, be obedient, and meet societal obligations
*   **Contribution**—To make a lasting impact on the world
*   **Cooperation**—To work collaboratively with others
*   **Courtesy**—To be considerate and polite toward others
*   **Creativity**—To have new and original ideas
*   **Dependability**—To be honest, reliable, and responsible
*   **Faithfulness**—To be loyal and trustworthy in relationships
*   **Family**—To create and sustain a happy, loving family
*   **Genuineness**—To act in a manner that is true to who I am
*   **God's Will**—To seek and obey the will of God

- **Growth**—To continue learning, changing, and evolving
- **Health**—To be physically well and healthy
- **Hedonism**—To simply enjoy myself and satisfy my desires
- **Helpfulness**—To be helpful to others
- **Humor**—To see the humorous side of myself and the world
- **Industry**—To work hard and well at my life tasks
- **Inner Peace**—To seek out and experience tranquility and serenity
- **Knowledge**—To learn and contribute valuable knowledge
- **Loving**—To give love to others
- **Mastery**—To be competent in my everyday activities
- **Order**—To have a life that is well-ordered and organized
- **Popularity**—To be well-liked by many people
- **Power**—To gain social status and prestige
- **Purpose**—To have meaning and direction in my life
- **Romance**—To have intense, exciting love in my life
- **Safety**—To be safe and secure
- **Security**—To protect loved ones, my community, and/or my nation
- **Self-control**—To be disciplined in my own actions
- **Self-esteem**—To feel good about myself
- **Self-sufficient**—To take care of myself without being dependent on others
- **Spirituality**—To grow and mature spiritually by connecting to things bigger than myself
- **Stability**—To have a life that stays relatively consistent
- **Stimulation**—To actively seek out adventure and create a life filled with novelty and variety
- **Tolerance**—To accept other people, as well as opinions and beliefs differing from my own
- **Tradition**—To respect and preserve the past and maintain order through tradition and customs
- **Universalism**—To create a sense of harmony among different people and preventing war and conflict; to create a sense of unity with nature and protecting it

- **Virtue**—To live a morally pure and excellent life
- **Wealth**—To have plenty of money
- Insert your own unlisted value:
- Insert your own unlisted value:

Once you select your top ten values, prioritize each one into the following categories:

1. Most important to me
2. Very important to me
3. Least important to me

Having clarified your most cherished values, you can keep them at the forefront of your attention, and it will be easier to determine which activities are worthwhile to commit to and be interested in. Returning to the "big-five" kinds of moments in our lives, we are regularly confronted with uninteresting or painful activities that are potentially hazardous to our health that we can quit (e.g., talking to a racist neighbor) or still feel compelled to do (e.g., calling your 85-year-old grandmother every week and enduring the 10-minute spiel about how you don't call enough).

Knowing your values makes it easier to make decisions as they offer us a direction to travel in. Enduring interest and passion—the hallmarks of a fulfilling life—start with moments of interest, that is, infusing daily activities with more delight, eliminating unnecessary activities that drain time and energy, and taking on new activities that have the potential to enhance our lives. But moments of interest are transformed when they tap into and reflect our values. It's easier to quit painful or uninteresting activities when we know that our cherished values conflict with their continuation. It's easier to justify ending a relationship if you know what values at your core are being disrespected or tread on. It's easier to begin relationships when you know which values you are more flexible about. Then there are values you don't identify with. Someone exemplifying them might expand your own world if you were to enter a relationship with them. Those unaccounted-for activities that are ruled out or rejected without even giving them a chance (see Chapter 4) can be reevaluated through the lens of our values. I cannot emphasize enough that you need to honor your own values, which may be different from any other person's. A fulfilling life stems from acting in ways that are aligned with our core identities. Values fill up space in the core.

However, your values can change, and your values might be different in various parts of your life. So, if you are unsatisfied with the moments in your relationships, think about what you value in relationships. The same goes for careers or how you spend your free time. If you can identify the exact area where you are unfulfilled, explore your values in that area and consider what to accept, what to transform, what to end, and what to begin anew.

The appendix of this book contains two other instruments that can help you evaluate your values and, more important, determine whether you are moving in the direction of your values and living a life infused with meaning. Complete each part of the Valued Living Questionnaire and the BULLI.

Finding the Right Home to Align Your Actions with Your Values

Doing something that is a good "fit" with your values, when you get to use your strengths on a regular basis, feels good. These positive feelings spill over into the activity itself, so you are more likely to be interested, devote greater effort, and persevere longer. It's a powerful tool for developing, strengthening, and maintaining interest and meaning. This notion of "fit" cannot be explained by how enjoyable a task is or how much of a positive mood people feel. When your values and strengths are linked to an activity, this sparks interest over and above feeling good.

We all know of situations that arm or disarm us, that seem to naturally fit with our strengths and values. Some of us are more people-oriented, which means we prefer to do things with someone rather than alone. Our strongest values may include helping others, sharing love, or being accepted. Ignore this element, and you are likely to find yourself in a number of situations where you function below your potential. If you possess a strong orientation to be around others, bring the social element into the forefront of moments where you felt disinterested in the past and activities you might have ruled out too soon ("off your radar"). Maybe it's just being in the same room with others; maybe it's cooking together; maybe it's working next to your closest friend at work so you can reward yourself after every hour of hard work with five minutes of playful conversation.

The same goes for attitudes. If you tend to be concerned about safety, responsibility, and preventing mistakes, you will be more interested and successful in a job where precision is essential.

If you were put in a situation where you needed to take risks and constantly adapt to changes, it would be a horrible fit. It would be strenuous and less motivating. For example, suppose you were in charge of focus groups that test-run jokes and stories for political speeches. This task requires a willingness to brainstorm ideas and potentially be laughed at and look foolish in hopes of nabbing a few bits of wisdom. This task is a great fit for a person focused on making gains (relishing positive outcomes even if they are few and far between), but horrible for a person focused on minimizing losses.

Let others know how you perform best. Your strengths are more likely to emerge when your values are in the forefront of your thinking and the environment you find yourself in (at work, socializing, and so on). When someone else is affected by how you perform, it makes sense to let them know the conditions that take advantage of your strengths and keep you optimally motivated. If the best fit for you runs against the norm, be willing to sell your ideas and perspective (based on the scientific evidence presented in this book) and suggest a probation period. For instance, for me to write, I need extremely loud music and minimal lighting. Picture a cavern with a small lamp, computer screen, and a massive set of speakers. Not exactly your typical, quiet, inviting professor's office. But without it, I can't function well. I know of businessmen who asked for a shower in their office because it helps them think when they're in a jam.

If your ideas work, then you get to operate your way; if they don't, you return to the status quo. It costs nothing, and if it works, others get to capitalize on the best version of you.

Strengthening Interests by Sharing Events with Others

For positive events to solidify and improve our lives, we need to hold on to them. Otherwise, just like yo-yo dieting where we keep returning to our original weight, we quickly end up back at our stable "happiness set-point."

Sharing details about ourselves allows us to feel close to other people. As a sign of satisfying relationships and a strategy for developing close relationships, appropriate self-disclosure is an essential social skill. I want to focus on another unheralded benefit. When we share our experiences with people who are important to us, the impact of their interest is profound. Laura, a woman in her thirties who enrolled in one of my college classes, explained the value of being fully present present when other people share their positive experiences.

I made a decision to try to keep the focus off of me when my friends tell me good things that happen to them. When one of my friends told me her husband was going to buy her a ridiculously expensive (and cute) Portuguese Water Dog, I was happy for her because she was bursting out of her skin when she showed me photo albums of him on the beach, in the grass, dressed up in a sweater, and on and on. I asked her what on earth a Portuguese Water Dog is, what makes it so special, and what her husband is trying to make amends for. I couldn't help saying, "I wish I had a man who would do that for me!" But I quickly returned to her bubbling excitement.

Even though we seem to always be talking and laughing for hours, there was something different about just focusing on her good news and nothing else. Asking for details made her excited and willing to share even more. She also seemed to be more excited about her news *after our conversation*, and less nervous about responsibilities, since she'd never had a dog before. I walked away feeling like I had done a good deed and this feeling lasted long after we left each other.

No matter how interested or curious we are in an activity, if sharing our experience with another person can strengthen this motivation, then we are better off for it. By sharing events that happen to us with other people, we create and strengthen our interest in that same event. When we share an activity with another person, the event becomes more prominent in the overcrowded files of memories in our brain. Sharing an experience with someone is like adding a bookmark, or flagging an experience so it can be easily revisited, remembered, and savored again.

Several studies show that the act of talking with others about an event effectively changes the way that we experience the actual event. We benefit from these positive memories because they are there for us when we feel down and need a mood boost. We can recall them readily and thereby feel better more easily and for longer periods of time. Also, we are reinvigorated and energized when similar events come our way in the future.

There's a crucial catch to sharing our experiences with others—the apparent interest of the person we share them with. If the other person is responsive and interested in what we have to say, then our own initial interest in an activity can be enhanced. We feel validated when our actions are viewed as worthy by others.

If the person we share our experience with appears to be distracted, dismissive, or disinterested, it is likely that we will come away feeling that our activity isn't very interesting after all. Robbing us of our original interest, our memory of the event will deteriorate, and it will affect our motivation when there is an opportunity to do something similar in the future. Doors to interests close rather quickly when we feel invalidated.

Capitalization is a term psychologists use to describe this social phenomenon. When we make a capitalization attempt, we are trying to deepen our interest and joy by sharing good news with someone else.

We can take the example of Andrew to illustrate this phenomenon. Andrew is a stay-at-home father who can't wait for his wife Jessica to return from her work as a hand surgeon so he can tell her that their one-year-old son, Sam, just took his first steps—seven Frankenstein-like foot stomps.

- An **active–constructive** response from Jessica might be, "Are you serious? That's amazing. Tell me all about it, walk me through everything. Where and when did it happen?"
- A **passive–constructive** response might be, "It sounds like you had a good day," followed by a smile and soft hand through Andrew's hair.
- An **active–destructive** response might be, "It's about time. Everyone else's kids started walking at 12 or 13 months. Why didn't you videotape it? You know it's important to me."
- A **passive–destructive** response might be, "That's great but tell me about it later. I had a really long day and need to take a nap. Oh, and remind me to tell you about the surgery I performed today."

There is only one good response and that's being active and constructive. An occasional passive response is normal and it's important to be true to your feelings and avoid fake smiles and laughter. What becomes problematic is when you consistently fail to share in the interest, excitement, and glory of the people you care about.

You might be surprised to know that how a partner responds to our positive events is more important to feeling valued, cared for, and understood than how a partner responds during difficult times. When partners are active and constructive when we share our interests, we experience greater well-being, our relationships are healthier, and we are more likely to maintain the relationship. When you share positive events, you are talking about your interests and strengths and there is something particularly important about having people validate what we care most about and what we are best at. Being there when things go right truly matters.

We can encourage and enhance constructive reactions to our experiences by putting the following guidelines into practice:

- Ask your partner if the timing is good before you share the event. You don't want to set yourself up for a failed attempt.

- Provide a story with a beginning, middle, and end. It's easier to get other people to listen and become interested in something that doesn't impact them directly if they can follow along as to why it is meaningful to you. Let them know why the event is important to you.

- Even boring events can be interesting to someone if they're told in a playful manner. We habituate quickly to the interesting parts of our job and find them to be dull. When we share them, other people often have that interested, exciting reaction we had when we first started. Also, self-deprecating humor, in small doses, can draw someone to you.

- Don't assume that your partner is dismissive and uncaring if they show their enthusiasm by sharing a similar event of their own. Triggering similar, memorable interests in others is normal.

- Take a close look at their nonverbal cues of attentiveness and interest (strong eye contact, nodding, smiling, saying "Great!" or reaching to touch you), as not everyone knows exactly what to ask and say and they may not talk as a sign of respect of your own enthusiasm.

- Be open to fresh, new perspectives that your partner might bring. The details that are boring, tedious, painful, and monotonous to you might be intriguing to them. Don't be defensive. If you are primarily concerned about appearing smart or right, you might lose out.

- Assume the best in your partners for a win-win situation. Remember, it is how you think they are responding that will influence your interest, so don't miss cues that show that they are really excited for you.

- Think about who you want to share a particular experience. Not all friends are equal. Each offers something unique, and it often has nothing to do with being a good listener. Choose wisely! Enlist the right people at the right time to bring more intense, frequent, and lasting interest into your life.

- Relationships are about reciprocity, and you are not going to get good responses unless you do the same for others. Thus, when your partner is trying/attempting to capitalize on an experience and looking for you to listen, practice the following:
- Listen actively and empathically to their account of the event.
- Mirror their interest and enthusiasm about the event at the moment as they are sharing it.
- Be genuine in your interest and enthusiasm. People pick up on fake smiles and laughter.
- Ask and probe for details about the event and focus on the positive. How did you feel when it happened? How did it happen? Tell me everything! After the story emerges, get into more specific questions to stretch out their interest.
- Without prompting, the next time you talk to them, bring up the event in conversation. Let them know that you listened and that what is meaningful to them is meaningful to you. It further solidifies their positive feelings and strengthens your relationship with them.

We cannot ignore the importance of other people in our lives, and we can strategically take advantage of these social connections to create and enhance our interests. Conversations serve as a medium through which others' reactions help construct and strengthen our own ideas about an activity as being valued, interesting, and worthwhile. What is important is that other people don't just change how we feel, they change our experience of the event, and this carries onward long after the conversation is over. Our interests can strengthen if our partner is engaged. If the conversation doesn't go smoothly, initially intense interests can quickly diminish.

Work As a Reservoir of Interest and Meaning

Having introduced a few key components for creating lasting interest, we need to give special attention to work because it comprises so much of our days, weeks, and lifetimes. Work is one of the central places where people can find a home and outlet for their lasting interests. For some, work is a reservoir of value and meaning.

If you are like me, when you were in grade school you took a career-aptitude test to discover if you were destined to become a heart surgeon or air conditioner repairman or don a toxic-waste protection suit to remove asbestos from abandoned crack houses. As silly as these tests are, the little pieces of information they provide may inspire us to look more closely at what we might do with our lives. It can kick-start the search for meaning. Of course, without the requisite desire to investigate yourself and to explore that new piece of information, the results you get from the test just get filed away. Free information and guidance without curiosity or interest is not enough.

I don't have to search my garage for the crumbling cardboard box where I keep my old report cards, love letters, and career-aptitude test scores. I remember exactly what the test results said—you should be an accountant and at a distant second, you should be a psychologist. At eleven years old, I didn't even know what a psychologist did other than smoke a pipe and listen while clients stretched out on a couch and babbled about their lives. I do remember taking the advice seriously and I spent several years on Wall Street playing with numbers before realizing that the mind, not money, might be my true calling.

A Job, a Career, a Calling

You might be one of the fortunate minority who spend dozens of hours each week getting paid to do work that feels like play (most of the time). Even if money and prestige are important to you, you find it personally satisfying to excel and do your best. You look forward to time away from work but also look forward to getting back to work. The sadness that starts on Sundays as the weekend comes to a close and the Blue Mondays that affect so many people, don't affect you.

If this does not describe your work life, you may think this idea of being passionate about what you do for a living is far-fetched and attained only by the most exceptional people. But as we saw with Fred at Grand Central Station, it's not. The same number of people who despise their work feel their work is a calling or purpose in life. This might be surprising because the people who feel their job is a curse tend to be louder and more vocal than the passionate workers.

Depending on the survey, 25%–33% of people experience their work as a passionate calling, another 25%–33% describe their work as a job, and another 25%–33% describe their work as a career. The type of work matters, as people doing "white-collar" or "professional, highly skilled" work are more likely to describe themselves as finding a passion or purpose.

According to a 2007 Gallup survey of thousands of people in more than 131 countries, the United States leads the way with 27% of workers feeling a strong emotional bond to their work, and 50% of Americans (compared to 70% worldwide) said their jobs gives them a chance to do what they do best every day. They get to use their strengths, they are very satisfied with their work, they are engaged in what they are doing, and they learn new things on a regular basis. Americans can take great pride at the number of people who gain so much from work. Yet these findings also mean that 7 out of every 10 workers don't feel a strong connection to what they do and 1 out of 2 workers don't get to use their strengths every day.

It's Just a Job (And Sometimes I Hate It)

Perhaps you agree with the following statements:

- I don't expect a lot from my work. It just provides me what I need to do the other more important things in my life.
- I often dread going to work on Monday morning.
- I think work is overrated when you consider what percentage of our lives we spend working as compared to enjoying life. I don't think much about work when I'm not there.

For this group of people, work is a necessary evil, and their jobs are nothing more than a vehicle to make money to support themselves and their families. For some, it's even worse. They count the minutes until the end of the day, and put their lives on hold until the weekend.

According to a recent Gallup poll, 25% of American workers felt like "screaming or shouting because of job stress," viewing it as a black hole of pain, boredom, and exploitation. Among workers younger than 30, 34% would quit their jobs if they could. In a study of 300 workers in Canada, from college professors to company managers to line workers, 23% didn't express a single bit of passion for what they do for a living. Nothing. That is, they didn't like their job, they didn't value what they did, and they wanted to spend as little time and energy as possible on the job.

If several of these qualities describe your attitude, then you have a job. If you are focused on the financial rewards of working more than pleasure or fulfillment, there is no reason to be ashamed. You might have meaningful outlets to put that money to good use. Plenty of people find true fulfillment outside the confines of their 9-to-5 existence. If you despise your work and are depressed and miserable all week, you would benefit from questioning why you are staying in your job. Figure out whether changing jobs or quitting is an option.

I Have a Career (And It Provides Enough Rewards to Be Satisfying and Meaningful)

Perhaps these statements capture how you feel about your work:

- While I enjoy what I do at work and am very good at it, I often feel like I'm tapped out and look elsewhere—my home, my spiritual life, my friends, my hobbies, my community service—to be fulfilled.
- The greatest experience at work is when I'm recognized by others for what I've accomplished.
- I will do whatever it takes to be successful at my work.

Many view their work as a springboard for much of what they want in life. Through their career, they can achieve all sorts of benefits including financial security, status, prestige, and an opportunity to express their talents and capabilities to work toward important goals. From these successes, they gain a foothold in enhancing their self-worth. This gives them a positive attitude about what they do, motivating them to put in more effort. By giving more to their work, they gain even more fulfillment. Thus, work is rewarding.

A number of people become doctors or lawyers or school teachers or get graduate degrees because it's what their parents want. Doing something that is valued by friends, family, bosses, and society makes them feel worthy. The desire to be popular and sit with the cool kids doesn't end in the teenage years.

If you have a career, you care about more than just money. Your main focus is success and accomplishment. This isn't to say you aren't immersed in tasks, and it doesn't mean you are lacking in pleasure and meaning. It's just that the work is a means to an end. It's often about the self-esteem boost you get from the external world—praise and recognition. Again, this is nothing to be ashamed about. One of the keys to a fulfilling life is being fully aware and receptive in the present moment and making progress toward meaningful goals. We get meaning from working toward important issues and feeling good about what we do and the type of person we are being and becoming.

I Found a Calling (And It Seems Like Work Is Play)

Perhaps this is how you think about your work:

- In some small way, I feel as if my work truly makes a difference in the world.
- There are moments when I think to myself, "If I won the lottery, I'd probably still be doing this work." I do what I do because I love it.
- I feel like my work allows me to show the "real me." My work challenges me, my work lets me use my strengths, and I tend to be absorbed in what I'm doing and lose track of time. These moments add up to something bigger that I get to contribute to and I am grateful for finding it.

In one study, around 33% of adults felt a calling or purpose best described their work. This is likely to be an overstatement because passionate workaholics—people who are out of control and obsessed with their commitment to work at the expense of almost everything else—are also likely to describe their work as a calling without getting any of the benefits.

Focusing on those with a healthy but not obsessive passion for their work, we uncover people who work hard because it is intrinsically rewarding. It's an essential part of their identities, and it serves as a renewable source of energy, pleasure, and meaning whenever they need it.

Think of Robin Williams' character as an inspiring teacher and mentor in *Dead Poet's Society*. Think of the neighborhood bartender who serves as friend and therapist. She knows her patrons love sagas, wounds, and favorite pastimes, and takes great pleasure in providing the emotional uplift or shoulder to lean on. Tips aren't what make the job rewarding; it's the connections to a diverse group of people.

Learning About Callings from Observing Those Fortunate to Find One

I recently had the pleasure of meeting Dr. Ray Fowler. He didn't know who I was, but I knew he was a distinguished psychologist who was interested in forensics, politics, and prison reform.

In the mid-1970s, "Dr. Ray" led a team of students who evaluated every one of the 4,000 prisoners in the Alabama correctional system. He wanted to improve the quality of care and security in this disorganized system. To do so, he created a thorough strategy for classifying prisoners according to their mental health, the threat they posed, and their potential to learn how to be healthy contributors to society. Recognizing that some prisoners could be reformed, he developed plans to treat, educate, and rehabilitate them. This ambitious undertaking changed the way that many prisons operated.

This is just one example of the projects he undertook over the course of 50 years, using scientific research to help people in the community.

Among other achievements, he served as president and later, as CEO of the American Psychological Association, the largest psychological association in the world.

When I met Ray and his wife in the summer of 2008, he talked about his retirement in 2002. Most of us think of retirement as the end of work. In fact, it happens to be the definition of the word *retirement*. Not for Ray. We met at a positive psychology conference in Croatia, where he had been meeting with people from Iceland, Israel, Australia, Greece, England, and other realms of the globe to create an international community of professionals focused on improving the psychological health of humanity (the same lofty goals you often hear uttered by Miss Universe candidates).

Ray goes above and beyond the call of duty for this passion, and he seems to do the same for everything else. Within 10 minutes of meeting, he told me that I should head north to Slovenia. He and his wife took off for a half-day trip through the Postojna Caves, one of the largest caverns in all of Europe, measuring 20,570 meters. He suggested that I hike a few kilometers and enjoy a ride on one of the rail carts that burrow through another few kilometers. His enthusiasm was contagious, illustrating how getting older and retiring can open doors to new adventures as opposed to serving as the end of an era.

"At 75, I am healthy and energetic, and I love what I'm doing. My workdays now [are] about as long as they have ever been, but not being office-bound gives me the freedom to be where I want to be including visits with our children and grandchildren who are scattered about the country. In spare moments, I am an avid bread baker and cook. My exercise program includes running, swimming, biking, hiking, and workouts at the gym. Unlike most people, I enjoy everything about traveling, even overseas flights and long airport layovers. I was elected president of the International Association of Applied Psychology in 2002, as part of a leadership cycle that continues until 2018, when I will be 87. I plan to serve out my term. Then—who knows—I may really retire."

Get this. If you do the math, When Ray steps down, it will be 16 years after the day he claimed to retire! Through his passion, you can see what it means to say that work is more than a job or career. Infused with meaning, his purpose is central to his life space and he isn't begging for vacation time. His purpose energizes him, creating room for family and long-standing passions for travel, cooking, wine, and triathlon workouts. Like Ray, we can expand the borders of our life and squeeze in more psychological and physical benefits when we find a match between our cherished values and our work.

Meaningful Work Benefits Many—Not One

Given this talk of work as calling, you would think it only benefits the individual. Viewing work as meaningful benefits other people too. If you lose interest in your work, costly company expenses add up quickly. It might start with a few a few sick days and "mental-health" days and perhaps an argument or two that knocks down morale a few notches. It might escalate to gossip and complaint sessions, and finally, turnover. For the individual, it might have been a lot less stressful to have picked work that was a better fit. For the organization, all of that salary and training goes for naught as they have to find someone new and start all over again.

Meaningful Work Comes in All Shapes and Sizes

If you are in a high-status field that requires a lot of education and training, or is highly respected by society, you are more likely to consider your work a calling. But this designation is not reserved for any particular occupation. Think about it. Somebody somewhere is lost in play doing what others are doing for pay. For example, compare going for a hike in the woods once a month versus leading scenic adventure expedition tours.

Many of us feel tinges of envy and awe when we hear about "left-of-center" careers that also pay the bills. We wonder why our high-school guidance counselors forgot to tell us that we could produce documentaries of crocodiles and hippos fighting for supremacy in the lakes of Africa. Who knew that you could practice medicine on dangerous expeditions, providing high-altitude treatment to mountain climbers in the Himalayas or treating explorers plagued by aquatic creatures that leap into unsuspecting urethras in the Amazon rain forest. Why aren't these job options presented as an alternative to being a teacher, accountant, or lawyer?

The kinds of lives led by people like Ray Fowler, "extreme" physicians, or various film directors, cab drivers, bartenders, and stay-at-home moms who found outlets for their strengths or ways to feel connected to things bigger than themselves can be ours for the taking, too. We can learn how to be mindful and we can practice working with our values to engage with life in the most meaningful possible ways. Our curiosity and our willingness to explore are not fixed.

Guidelines for Infusing Work with More Meaning

Is your work just a way to earn money, or is it a path to expressing your interests and strengths, adding fulfillment to your life? I want to emphasize the importance of being honest. You can't understand the sources of meaning in your life and what to accept and what to change unless you are willing to suspend judgment and be receptive to seeing things as they are. Work is only part of the picture, and if it feeds the other, more important sources of meaning in your life, then this is excellent. But it's silly to use a single category to characterize your work, as things change from moment to moment. Some days your work will feel like a job and other days, it will feel like a career or calling.

The curious explorer within you can identify what is going on in the moment so you can recognize the rewards when they arise, accept the things you can't change, and trigger your engagement when you need a surge of energy, interest, and excitement. Here are some specific guidelines for infusing your work with meaning and passion:

1. Discover Your Strengths

Seeking, using, and capitalizing on the strengths you possess is unlikely to happen without some intense self-exploration and discovery. You gain little by having a general awareness that you are good with people, good with your hands, or good with creative challenges. Get into the details. Are you compassionate? Are you generous? Are you forgiving? Are you resilient when the cyclone of stress hits? Are you courageous and able to move forward despite your fears?

If you have trouble pinpointing your strengths, think about your accomplishments and proud moments—your best conversations, your best efforts, your best performance, and the times you beat the odds in difficult and challenging situations. When are you at your best? What are the things that energize you? Pay close attention to what you gravitate to in your free time. What books do you enjoy reading? What activities rejuvenate you? What challenges you? What do you look forward to doing? When do you feel the most authentic or effortlessly yourself? By analyzing these experiences, you will see your strengths and passions emerge. Give your strengths a name and with this new vocabulary, start to describe yourself in positive terms.

Rather than the usual ice-breaker of sharing names and favorite foods, on the first day of class I ask my students to describe an event in their lives that characterizes who they are at their very best. I don't care how long it takes for them to finish the story.

What happens after the students go around the room? At first, people feel uncomfortable because they are not used to sharing their strengths; they are taught at an early age to be humble. But then something else begins to happen. They feel a strong sense of excitement, awe, and appreciation of all the great people surrounding them. They feel a strong connection to these strangers because they learned about this core essence of who they are and it becomes a quick segue to talking to them after class. This one exercise lets everyone know that people whom you least suspect have profound strengths. Imagine what this could do in the workplace. Imagine how much better teams could perform, how sensible it is to use people's greatest abilities, and how social bonds can be quickly created if everyone learns something profound and positive about everyone else.

If you have difficulty thinking about your strengths on your own, ask a close friend, family member, manager, or coworker for a story about a time when they saw you at your very best. In fact, you might want to reach out to others anyway to learn about strengths you possess that you might be dismissing or neglecting.

2. Bring Your Strengths into the Workplace

You will be more motivated, perform better, and feel a stronger connection if you can bring what you care about into your work as opposed to waiting until 5 o'clock or the weekend rolls around. This requires awareness and experimenting but the effort is worth it.

The fact is, the best teams are made of people who stand up for their strengths and volunteer them to the team. Thus, by knowing your strengths and wielding them at work, you also will create a better group, which is yet another thing to be proud of and draw meaning from.

In a Gallup survey, employees were asked about what they most regularly talk about with their managers. The results showed that 36% said it was about fixing weaknesses, 24% said it was about building strengths, and 40% said they don't talk about these things with them. Don't wait for your managers to start the conversation, let them know what you can do and what you think your strengths are so that you can search for the meaning you want, find it, and add it to your identity.

3. Find a Secure Base

When you feel safe and secure in your position, you can be an engaged and curious explorer, taking risks and being creative and innovative. When you don't feel safe, you rely on what doesn't lead to mistakes and problems. When you start worrying about making mistakes and failing, you waste energy trying to control circumstances and what will happen in the future instead of being focused on the tasks at hand. This slows down your thinking and hampers your creativity, and everyone loses.

4. Take Action. The Only Thing Worse Than Doing Something Is Doing Nothing

Variety and challenge keep you attentive and interested in your work. If you never get absorbed and lost in what you are doing, you are not being challenged. Mix things up. As soon as you master something, appreciate your accomplishment and then ask for new opportunities that require more and more skill. It's just like working out at the gym or playing a video game. Our body actually shuts down once it realizes that it's not being challenged. Remember that your body is designed to conserve energy because all it cares about is survival. Don't fall prey to your lazy mind. When you go on autopilot, you lose out on the chance for pleasure, engagement, and meaning. Being challenged on a daily basis and feeling supported by others are the twin gatekeepers to helping you find and create meaning at work.

5. Focus on People as a Source of Support and Meaning

People are often the first reason why we rule out work as a source of meaning. When we feel the support and connections to the people around us we naturally experience a greater sense of meaning. When we leave our jobs, whether by mentally tuning out or actually quitting, it's often because we experience a lack of recognition about our strengths or a lack of consideration of what interests us. The best thing we can do is be assertive and express what our strengths and interests are to other people. Focus on ideas and solutions and not just what isn't working. If you think your manager will oppose change, let them see how your attitude, efforts, and performance improve. After all, this is the bottom line they are focused on. You will be surprised how flexible organizations can be when you "sell yourself" as being underutilized and wanting to be more motivated and productive. In this book are some ideas for how to do it. By bringing people into your corner, you will gain strong social connections that will spill over into a stronger connection with your work. All of this sets the stage to extract greater meaning on a regular basis.

6. Find a Link between Your Identity and Both Your Work and the Work Team

Passion toward work is inevitable whenever we feel as if our real self is being exposed. When we feel that what we do and how we do it is important, we feel a personal connection. This connection doesn't have to be totally individualistic to be meaningful. The connection can be a collective one as well. This is the exact mindset of many people from Eastern cultures who don't separate themselves from the group they belong to. When their group triumphs, they triumph. When their group commands respect, they feel respected. We can learn a lot from this collective sense of belonging. Meaning derives from making sense of where we fit in. We can enhance our sense of meaning by remembering to be open to and curious about the people around us and, by observing just how many groups we are a part of and important to.

7. Become More Curious by Remembering to Be Curious

We control only our thoughts and behavior; we have almost no control over other people and situations. We often think we have to do new things to be interested and engaged in daily tasks. That is one path to fulfillment, but another is to find the new. Go into work with the mindset of looking for the two or three things that are novel and three things that can challenge you instead of relying on what is familiar and known. What is novel about your coworkers' attitude? What is novel about how your body is working? What is novel about the assignments in front of you? What challenge can you bring to your meetings, tasks, time-management, or relationships this day or week?

Treat your workplace like a research laboratory. Even if you can't lessen the number of seemingly mindless tasks that need to be done, you can explore, experiment, and discover during the day. In all of these instances, you don't sit and wait for things to happen. You are aware of what is around, you seek things out, and you play. It's strange that we put so little emphasis on fun and so much emphasis on how much money we make. It's strange that people who are trying to hire new employees put so much emphasis on education and training and so little emphasis on whether they are interested and excited.

Your work doesn't have to be your primary source of meaning, but if you do work full-time, you can expect to spend about 45% of your time at the heels of your occupation. Why not decide to do something that creates more energy than it steals? If you don't enjoy what you are doing, ask yourself whether you can change your perspective. If you view your work as serious, painful drudgery, it will bring down your mood. If you view your work as fun, challenging, and an opportunity to grow as a person, it will lift your spirits and change your outlook. If you can't change your perspective, if there is nothing that entertains or interests you, ask yourself whether you should be doing something different with this large chunk of your life on this planet. You can find purpose at any stage of life. There's no need to wait to seize what you want in life. Your life goes on, and it's up to you to realize that your curiosity can alter, accept, or appreciate that trajectory at any time.

Switching Perspectives: Strengthening the Interests of Others

Walking into a typical statistics class is unlikely to change your opinion about the subject if you are expecting it to be hard and worried that you are going to understand little. You have to believe the payoff is worth the effort, or you are going to move on. To develop sustainable interest and engagement, the value of the subject needs to be clear and personal.

On April 25, 2008, more than 4,000 high school students poured through the gates of Six Flags America for Physics Day. Is there a better way than a ride on a roller coaster for teachers to illustrate the gravitational pull of objects toward Earth and G force? For many kids, this could be a pivotal moment for them, when they realize that physics can help them start to unravel some of the mysteries that surround them.

When eminent artists, scientists, politicians, business people, and explorers of the twentieth century were asked by Mihaly Csikszentmihalyi, a psychologist interested in curiosity and creativity, to reflect on their life choices, many of them recall an intriguing experience or an illuminating moment that became the starting point of a life-long quest. This illuminating moment was often fueled by a mentor who recognized their curiosity and kept them stimulated enough to build on it. Without capitalizing on those moments to trigger and sustain these interests, who knows what insights the world might have missed.

There are strategies for helping others capitalize on their interests and strengths. When we tap into the power of curiosity, we find new and highly effective approaches for motivating others. The following examples are just that, examples. These strategies can be applied to nearly any situation in which someone is under your watch (managers, parents, teachers, and coaches). They can also provide insight into the nutrients you need in your own life to strengthen your own enduring interests.

1. Visualize the big picture and share it with people. If you manage workers in a large company, allow employees to see the big picture and how their work contributes to it. Few people are going to be intensely motivated if they are working in a bottling plant without being let in on what products are being created, what new ones are being planned, and how their actions contribute to increases in sales or prestige. Give employees a taste of the product, ask them for ideas, and show the link between their work and important company outcomes.

Sure, you should reward employees in proportion to how sales move, but knowing that they are doing something meaningful for the company is equally important. Give them time and allow them to provide input.

The same goes for when you know there are going to be transitions and changes that will affect other people. Whether you are thinking about moving to a new city and have yet to tell your kids about the possibility of a major change in their lives, thinking about upgrading the technology in your corporation and have yet to tell employees how it will affect their responsibilities—the sooner the better. What researchers find is that the earlier that people are brought into the change process and the more input they have in the process, including sharing thoughts and ideas, the more open they become to change. This is true even if it means that their lives are going to be shifted or reorganized. By increasing their openness, you increase their interest and commitment to the process. Transparency about our hopes, dreams, fears, and plans can enhance the interest of people we care about, and their positive feelings are often contagious, thus intensifying our own interest.

2. Reward people beyond money. Money can undermine initial, natural interest in a task. What scientists have found is that when people do an activity because they are interested in it and then you pay them for the same activity, their motivation deteriorates and they are less likely to do much on their own without the money.

A business manager I interviewed noticed that someone had planted new flowers in front of the office building and seemed to be tending to them on a regular basis. Upon discovering who the person was and expressing gratitude, she decided to give the woman a weekly bonus just as she would for a gardener. The result? A gesture of love became another work assignment. In all likelihood, the flower-loving employee will start to think less about the plants and more about doing well to maintain this added income. And if she stopped being paid, what we know is that she is likely to stop caring for the garden. Money is not meaningful on its own, so it makes sense to skip the money and focus on what you want to enhance—interest and meaning for the task at hand. I am not criticizing this particular manager's kind intention; I am pointing out that money often doesn't provide the incentives we want.

Create rewards with personal meaning attached to them. Discover what that is by learning about the values and interests of the person you are trying to motivate (employee, student, child, athlete). Two things happen. First, the person feels cared for and valued because you are showing interest in learning about them. This will increase their commitment to what you care about. Second, when rewards are linked to their strengths, values, and performance, you create a positive memory that is forever linked to the activity.

If you have an employee who is incredible at keeping work morale high and makes the workplace more enjoyable and everyone more productive, give out a citizenship award. But don't stop with a plaque; give the employee a gift that acknowledges a particular interest they have outside of the workplace, say, a weekend getaway to a favorite place or tickets to a special event. If they get money, they will use it or save it and probably consider it no more than an extension of their salary, and it will not have an iota of lasting meaning. For the same amount of money, a personalized event can connect people to tasks and organizations, enhancing interest.

As a caveat, if we are still in the early phases of getting someone interested, don't worry about any kind of reward. External motivation is not going to be enough. In the beginning, it's all about exposing people to something new and ensuring that the experience is positive, rewarding, and enriching—get that door open (see Chapter 4). What I am referring to here is rewarding people who are already interested. Our goal is not to trigger interest, it's to deepen and solidify it.

3. Ask people thought-provoking questions. Given the choice of anyone in the world, who would you want as a dinner guest and why? If money was no object, what would want to do with the rest of your life? Find out what they are interested in and what they are passionate about. These interests can be used to construct examples or metaphors that help them discover why a particular task could be valuable to them.

5. Listen, accept, and avoid confrontation. Let them do the talking and follow up on the breadcrumbs they offer. Save and reintroduce this information in future conversations to demonstrate that you listen and pay attention. Accept their interests and passions. If you don't like them, remember that they might change over time. For example, if you are a parent, your teenager's current obsession with poker can help develop keen skills of reading other people and thinking about statistics, which are essential to being an excellent scientist, detective, doctor, business executive, or teacher (just to name a few). Attempting to persuade them to do things that don't interest them will only cause them to feel criticized, become defensive, and withdraw emotionally, preventing access to anything else they are thinking about. Be open, constantly exploring and validating what they care about.

6. Create bridges. Encourage people to find connections between things they care about and things that they have to do but aren't particularly interested in doing. Consider parents and teachers who work with kids. Since most kids love their computers, cell phones, and whatever Jetson-like gadget is coming next, it's easy to draw connections to physics. Children who are interested in sports, from skateboarding and rock climbing to playing soccer, may be surprised to learn that biology and physics are used by the best professional athletes to attain peak performance. Besides careers, there are the tasks of everyday life—figuring out a waiter's tip, measuring ingredients when cooking, and understanding how the cost of gas affects where the family goes on vacation. It's important to help people find value and meaning as opposed to shoving ideas down their throats. If they have trouble seeing connections, you should be ready to get them started and guide them with prompts and questions. Be an example and show them how your own life is influenced by the same topics and activities you are trying to get them interested in.

Let's return to Bo Kimble's basketball coach and his willingness to accept a few lost points so that his athlete's values would be recognized and appreciated. Whether you are a manager, supervisor, or parent, you need to figure out what you want from others. If the goal is for them to be maximally interested, engaged, and productive with high morale, then you need to be flexible and allow them freedom to find the fit that works best with their values and personality. It might be different from what others are doing, but that shouldn't be a problem if productivity, creativity, inspiration, energy, and morale are high.

If the person under your guidance wants to take an unusual approach that is personally meaningful to them, ask them to sell it to you (why will this help you become more interested? more productive?). If you are persuaded, accept it for a probationary period. The worst that could happen is that they try their hardest and end up as disinterested as before. Either way, you are viewed as a caring open-minded person and it costs nothing. If you fail to experiment with new accommodations, you risk underestimating and underutilizing people's strengths.

7. Persist. In the early phases of these discussions, your primary goal is to open the communication channels. Interests wax and wane, and the entire process of creating meaning and value can be lengthy. Take pride in small gains as they figure out what does and doesn't fit with their values, and whether they need to change their perspective or values. We aren't interested in getting them to merely feel good; we are helping them acquire a sense of personal ownership in their life, especially the parts where they feel controlled by other people (administrative tasks, household chores, mandatory courses). Stumbles, set-backs, awkward moments, and feeling uncomfortable are part of the process. Persisting in the face of these obstacles is important.

Owning Moments

We become accustomed to long-standing routines far too easily as we continue along the treadmill like proverbial lab rats, walking, running, eating, working, going to sleep, waking up, and doing it all again. The old adage to "know thyself" is useful, but far too often this is the only advice we get, leaving us to ask "okay, but now what?" Thankfully, a great deal of science has emerged to suggest several "interest enhancing" strategies to transform moments and create lasting interests. As shown in the last two chapters, with effort and intention, we can reclaim and energize seemingly mundane, boring, tedious, and repetitive day-to-day moments and ensure that the ultimate currency of energy is well spent. By doing so, we create value and meaning while navigating the terrain of our continually evolving, unique niche in this world.

6 The Rewards of Relationships
Infusing Energy and Passion into Social Interactions

They'd been married for three months now, and everything about her was a novelty and a revelation, right down to the way she stepped into her jeans in the morning or pouted over a saucepan of ratatouille, a thin strip of green pepper disappearing between her lips while the steam rose witchily in her hair.

—T. C. Boyle

The opposite of love is not hate; it's indifference.

—Elie Wiesel
(

I graduated from college with a focus on investment banking and went to work on the floor of the New York Stock Exchange. It was the job I seemed born into. My grandmother was one of the first women to crack the glass ceiling in the field of finance, and I had several cousins who earned a living on the floor and subscribed to Gordon Gekko's "greed-is-good" lifestyle.

A few years into my money-making career, I had a revelation while hanging out with a few close friends on a golf course in the wee hours of the morning. Nearly every week that summer, we would illegally hop a fence with our lawn chairs and beers, and talk until the sun rose. Each of us had girlfriends and weekend party invitations, but nothing brought us more pleasure than being out in nature (golf courses being the extent of nature in the suburbs), sharing our philosophies, and laughing while the rest of the world slept.

During one of those nights, I was whining about the monotony of the stock market and how my sanity was being saved by stimulating books that awaited me at home. The books we read bled into our conversations. I was talking about Philip Dick's surreal, pessimistic vision of what the United States might look like if it had lost WWII to Germany and Japan. Concentration camps flourished and everyone, from corporate executives to gas station attendants, would forgo logic to make decisions by turning to the ancient oracle known as the I Ching. Other topics simmered within me, from the thin line separating genius and madness to scientific discoveries about love and creativity. My friend Dave scrunched toward the edge of his chair, looked straight into my eyes, and said, "I don't get it. If you're so interested in these topics, why don't you go into psychology? Feed your starving brain and save the stock investments for your free time. Flip your worlds." And that was it. Maybe it was the susceptibility of that field-of-dreams moment (fueled by alcohol and sleep deprivation), but at 4 A.M. that morning I decided to become a scientist. But first I needed the right mentor.

In the late 1960s, Arthur Aron was a graduate student getting his doctorate in social psychology at Berkeley amid the backdrop of radical social changes. Elaine was a student in his T-group (T=trust). In this emotionally charged atmosphere, the two of them were at odds throughout this group process course. Elaine had extensive leadership experience and thought Art was running these groups wrong. When the semester ended, constraints lifted, and, as Art says, "I remember that very day. We walked onto the campus quad, looked at each other, and though we had been fighting constantly up to this point, we kissed. It was very intense and basically we have been living together ever since."

As budding scientists falling in love, they wondered why nobody was studying this profound human experience. So began a shared career dedicated to understanding what it means to be close to another human being, and the circumstances and meanings of intimate relationships.

My path crossed with Art's soon after I left Wall Street. I was searching for an entry-level research job in preparation for graduate-school applications. I stumbled upon Art's lab, and I vaguely remember the phone conversation that unveiled the start to my career. Art was looking at how couples in committed relationships re-create the passion of the first few months of getting to know each other when the emotional drama peaks. As an unpaid assistant, I burned through my meager savings, and immersed myself in the science of friendships and romances.

Expanding the Self by Entering Relationships

My personal story continues with one more twist, which brings the scientific research home—literally. Partway into my internship, I was single and dating when my friend Martin invited me to a rooftop party in New York City. When I arrived, a smiling, six-foot-tall, well-sculpted woman opened the door. She was the hostess, and I was smitten. Pursuing her was an exciting challenge. The apartment and rooftop deck were filled to maximum capacity. One minute, she was dancing on a table; the next, she was getting a cluster of guests laughing, mixing drinks, or checking out the view overlooking the city. I found her on a couch later in the night and took a place next to her.

She asked me what I did, and I told her that I was studying how couples maintain passion in long-term relationships. When I saw how intrigued she was, I realized that my new line of work had given me a killer pick-up line.

She was a yoga teacher and more interested in intuition than science, but she seemed eager to learn more and plied me with questions. Can you really study passion and love? What have you discovered? Our conversation continued for hours. While she absorbed my love for science, I absorbed the details of her yoga training, how the philosophy behind it is often lost, and the ways of calming the mind through body movements. Six months later, we married. From the beginning, we had different values and interests, but instead of repelling each other, we found harmony in giving each other access to these independent worlds. As a couple, each of us grew into larger, more complex beings. Over time, as we reached a saturation point of learning the basic details about each other, as happens for all couples, we have sought new ways to grow and expand who we are.

This sometimes effortless but often challenging and stressful process is at the heart of creating and maintaining meaningful relationships. It is my contention—borne out by research and clinical experience—that relationship interest, excitement, and commitment are skills that can be learned and applied in all kinds of circumstances and connections.

Psychologists, including myself and my research team, have recently found that curiosity, novelty, and uncertainty provide clues to why we enter relationships, why we are attracted to certain people and not others, how we can enliven conversations to be more interesting and enjoyable, why relationships falter over time, and ways that passionate intensity can be extended over time. This early research, and this chapter, focus on romantic relationships and the magic of falling and staying in love, but the findings and insights are relevant to all kinds of social activity from encounters with strangers and conversations with coworkers to our most important connections with parents, children, friends, and with all the people we want to inspire and help grow.

In this chapter you will gain insights, some counterintuitive, about particular mindsets and behaviors that allow relationships to thrive. Instead of waiting for breakdowns and trying to fix them, you will learn about adding positive energy, intrigue, and meaning to the time you spend with other people. Experimenting with these insights might lead to huge dividends in your own life and to the lives of the people you care about most.

Why Relationships Work for Us

If there is one characteristic that separates very happy people—the ones who laugh and hug strangers while kamikaze pigeons shit in their hair—from the rest of society, it's the quality of their relationships. Very happy people have satisfying and meaningful connections with others.

Robert Biswas-Diener, a happiness researcher, interviewed the homeless in Fresno, California and prostitutes in the slums of Calcutta, India. Although people in Fresno had greater access to food, shelter, and free health care, the prostitutes in Calcutta were happier. Why? Returning to the notion that there is more to a fulfilling life than happiness, Robert concludes, "while the poor of Calcutta do not lead enviable lives, they do lead meaningful lives." Prostitutes in Calcutta possess stronger social ties and have regular contact with family members, allowing them to cope more effectively with the destitution, decay, and degradation. Strong social relationships provide meaning and moments of bliss in the surprisingly safe, low-pressure center of the cyclone, where theoretically, you can remain protected and unscathed even when surrounded by reams of destruction.

Strong social relationships create meaning and well-being. Daily social interactions are equally important as the building blocks for creating good relationships; even when they don't lead to relationships, they are the springboards for the best and worst moments in our lives. Feeling rejected and hurt by other people is often the cause of intense, lasting pain. Alternately, feeling accepted and cared for by other people inspires the greatest comforts.

Although we are no longer solely guided by that ancient, reptilian part of our brain that tells us to "eat, %$#@, and kill" so our genes will survive another day, our brains still push us to explore and reward us for seeking out the new. These are the same strategies that led our ancestors to find new sources of food, sexual partners, and creative ways to defeat enemies and increase social status.

A close look at modern life shows that we spend an enormous amount of time and effort to increase our skills, knowledge, experiences, power, and social resources to make inroads toward important goals. Sure, we are motivated by the pull toward safety and seek to avoid danger, but we also possess a fundamental motivation to expand and grow as human beings. Not only does growth feel good, the wisdom and strength of knowing more and increasing our skills gives us an advantage at surviving and thriving in an uncertain world.

Perhaps the best way to expand and grow is to enter into close relationships. For example, when two people first enter a romantic relationship, it begins with a whirlwind of exhilaration. Being curious and wanting to know about our partner is important and by acting on these desires we develop a meaningful relationship by "including the other in the self." This is typically done by spending time and sharing experiences together. We mutually disclose the details of who we are now and who we were in the past, and we remain open and respectful of how we differ from each other and alert to common ideas and interests. From these exchanges, we acquire the knowledge, insights, abilities, and possessions accumulated by our partner over their lifetime (and they do the same). This is a period of rapid self-expansion as each of us takes turns sharing and absorbing the other into our identity (or sense of self).

You see a similar pattern in the friendships of young girls who spend hours and hours talking, emailing, texting, and carving out large chunks of time to be together. You also see this pattern of passionate self-expansion in any friendship where people are becoming confidants to each other, though with less intensity in men's friendships.

Especially in beginning romantic relationships, novelty is in abundance and it's arousing and exciting. When our partner is on a quest to get to know us, we feel desirable and worthwhile. Someone wants to unravel the superficial layers of our personality, penetrating deep into our hopes, fears, aspirations, strengths, weaknesses, likes, and dislikes. When we compare ourselves to how we lived before the relationship, we recognize how much we've grown in a matter of weeks or months and sometimes find it hard to imagine who we were or how we lived without this other person in our life.

I spoke to a middle-aged man about his dating partner of three months. Today, his social calendar is packed with dinner plans, parties, and tickets to music events in out-of-the-way clubs. He confided, "If only some of these characters saw me back then, planted in front of a computer eight hours a day at work only to come home, pop in a Pop Tart, and spend the night playing video games. I still have my old self and now, with some improvements added to it."

Expanding by being in a close relationship is similar to a shared bank account where each person has access to the other's money. But the exchange goes beyond the tangible. Yes, it includes being able to share material possessions (I can use your stuff). But it also includes knowledge and wisdom (I can use your brain) and an expanding social world (I can spend time with and count on the people you know).

When we are in a largely expansive relationship, besides feeling closer to our partner, we actually become linked to them—their qualities become part of us. If my partner enjoys fishing and we regularly spend time together on a boat, then simply from being together, I increase my knowledge of weather and currents, my abilities to bait a line, and new topics of conversation start coming up when others ask what I do and who I am. Our partner's world is woven into our own like a beautiful tapestry.

This explains why we can get so protective when someone criticizes our friends, family members, and romantic partners, and in turn, why we feel a surge of pride when they are complimented. In a weird way, the person is criticizing or complimenting us. It also explains why "breaking up" is the ideal description of the loss of a relationship. The material possessions, knowledge, insight, philosophies, values, friends, and unique perspectives of the world that each partner brought into the relationship to be shared are no longer available to each party after it ends. Each person expanded by being in the relationship. After breaking up, they become a fraction of that union.

Over and over we see that the best relationships—romantic and otherwise—involve two unique partners who recognize and act on the value of knowing each other through an ongoing process of sharing, exploring, observing, respecting, integrating, and, depending on the situation, accepting or changing. In general, we feel closer and more satisfied in relationships that provide ample opportunities for these self-expansion opportunities. Construing relationships as the ultimate opportunity for growth allows us to understand why some relationships satisfy us and others don't.

Self-Expansion, Relationships, and the Brain

If people do grow when they begin close relationships, then it should be visible. And it is. After falling in love, people use a greater variety of qualities to describe themselves than they did before. People who fall in love expand their definition of who they are as a person; their sense of self expands.

When married partners talk about their relationship to interviewers, they use the term *we* more often than the word *I* compared with lovers in the first few months of being together. A level of "oneness" starts to take place as partners mutually share their lives and grow together. This is a good thing, as one study found that lovers who shared more of their lives with their partner and integrated parts of their partner's world into their own (self-expansion) are more likely to still be together three months later. They also show greater intimacy, commitment, passion, and a willingness to sacrifice their own needs for their partners.

This process of self-expansion in close relationships is not limited to romances. In a social interactions, rapid self-expansion during conversations (such as juicy information that feeds our curiosity) leads to the experience of greater positive feelings and affection. We remember this positive moment and are more willing to spend time with these conversational partners again in the future.

Some fascinating research shows that as a result of being in a lasting same-sex friendship we consider many of our friend's qualities, interests, and values to be part of us. Often, it can be confusing to clearly define the boundaries of who we were before we met them—before they started influencing and changing us. What I find to be fascinating is that this process is visible in the brain.

In one study, researchers asked subjects to give the name of a close, same-sex friend. Next, subjects entered a brain scanner (functional MRI) and for several minutes they heard a number of different names including their own, the name of their friend, and common names from their culture (names they are likely to have heard and that might signal someone they know, but not anyone in their inner circle of confidants).

The researchers knew from other studies that certain brain regions light up when we think about the self. This makes sense. I would have a hard time getting food in my mouth or answering someone's question if the line was blurred between where I start and everything else begins. Other brain regions are activated when we think about other people. These findings offer tantalizing ideas about what happens to our brains when we meld our lives with someone else. What scientists wanted to know is if I'm in a close relationship and include my close friend in my sense of self, does the part of my brain that lights up when I think about me light up when I think about them?

Yes, according to what these neuroscientists found. For people in close relationships, when they heard the name of their close friend, the brain region linked to the self lit up. It didn't happen when they heard familiar names.

The brain still separates the self and other people, but it makes a special exception for friends, lovers, and family members who are intimately intertwined with our own lives. When you form close bonds and trade access to your innermost self with close friends, these relationships get imprinted in your brain. Your brain gets rewired, adding new neural connections to include the people most important to you. The brain literally changes as a result of experiences.

Caveat Emptor: When the Need for Safety Trumps the Need to Explore

Unfortunately, not all relationships and conversations satisfy our curiosity and provide the stimulation we desire. Among the reasons why a relationship can be unhealthy or undesirable is an imbalance between the need to feel safe and secure and the motivation to grow and learn. Impending danger can squelch attempts to act on interest and intrigue. Perhaps the person doesn't respect and care for your well-being and might even be hurting it, whether through neglect, condescension, contempt, or even aggressive and violent behavior. Instead of trying to grow and enhance a particular relationship, our motivation shifts to avoiding pain, anxiety, and misery in order to protect ourselves.

Danger is a useful spark for triggering excitement and interest, but it is problematic for sharing worlds. Think of women who show a preference for "bad boys." It sounds good in the short-term, but there is a trade-off. True intimacy and growth stagnate when we need to be vigilant about being hurt. The evidence is clear that when our partners view our feelings and our relationship as a game to be played, intimacy suffers.

There is one exception. When couples are similar in their craving for thrills and adventure and fears of being bored, they feel closer, have better sex, and are more likely to stay together. Two thrill seekers make for a good couple. They understand each other's needs and provide enough stimulation to satisfy each other.

All Relationships Start Somewhere: Sparking Excitement to Attract and Be Attractive to Other People

All relationships start somewhere—in chance meetings, ongoing social interactions, and friendly connections that percolate over time. All of these situations are opportunities that can enhance our lives by expanding our horizons. Creating and seizing those opportunities is a shared dance.

On the one hand, we are deciding whether someone is worthy of our time and energy when compared to all the other people we can be with at any given moment. On the other hand, other people are making the same decisions about us.

From recent scientific discoveries, we can walk through some of the steps of this dance that triggers interest and excitement in others and ourselves.

Curiosity Acts as a Positive Spark in All Relationships

We know that relationships are essential to well-being. And we're learning that curiosity propels us to connect with others so we can expand and grow. But we know less about how being a curious person is beneficial in creating healthy, meaningful relationships. Does it feel good to share your life with a very curious person, and if so, what is it that's desirable about them? I have devoted several years to conducting research to answer these questions.

If being very curious improves social interactions, then it should be apparent to lovers, friends, and family. It should also be apparent to researchers who observe people talking behind a one-way mirror. I conducted some of the first studies on the social consequences of being curious. Strangers were brought into our lab to spend five minutes getting acquainted. Before these interactions, participants filled out questionnaires so we could rank them in order from least to most curious. What we found was that the more curious a person was, the more they enjoyed interacting with strangers and the closer they felt to them by the end. But the positive experience wasn't just "in their heads." The partners of the most curious people felt the same way, too.

In looking for the traits that led to these beneficial exchanges, we found that curious people treat their partners as vast unknown territories to be explored, asking lots of questions, as they continually penetrate deeper into new terrain. The people who talked to very curious people appeared to enjoy being in the spotlight of this undivided attention.

It feels good when someone is interested in you. It makes you feel desirable, causing you to be more interested in and feel close to the person showering attention on you. Perhaps best phrased by Dale Carnegie in his 1936 classic *How to Win Friends and Influence People,* "you can't catch fish with strawberries and cream."

Along with scientists at the University of California at Davis, we took another step toward understanding what exactly curious people do that makes them desirable to be around. We wanted to look at how curiosity affected long-lasting relationships as well as new ones. There were two parts to this study. Once again, participants interacted with strangers for a few minutes to determine what type of impression they made. We also conducted interviews with participants' college friends, "best" friend, and parents about their strengths, weaknesses, and how they typically act. By combining these perspectives, we created a picture of what it feels like to spend time with very curious people.

After a mere five minutes of getting to know them, strangers said that curious explorers were different than less curious conversation partners in several ways. Among other things, they tend to exhibit the following traits:

- Display a wide range of interests
- Highly enthusiastic, energetic, and animated
- Talkative
- Dominant in conversation
- Express their ideas well
- Say and do interesting things (they were intriguing and sometimes, surprising)
- Appear relaxed and comfortable
- Completely attentive and engaged in conversation

What is remarkable is that these strangers picked up nearly the same qualities adored by the college friends, best friends, and parents of these same very curious people. For instance, here are the qualities that most distinguished very curious people from less curious people:

- Displayed a wide range of interests
- Easily volunteer information about themselves (including provocative topics such as sex and religion)
- Dominant in conversation
- Highly intelligent
- Express their ideas well
- Funny, often the first to be playful
- Appear relaxed and comfortable
- More open and less judgmental

Apparently, curious people don't hold back with people they just meet. We know they are authentic and acting like themselves because strangers recognize many of the same qualities pointed out by long-term friends and family.

Very curious people ask lots of questions, but it's not a one-sided exploration. They reciprocate by sharing intimate information about themselves. Besides the intense wonder and questioning, they express their ideas well, are attentive and engaged, and open to whatever arises. This is important because intimacy and closeness only happen when both people want to learn about the other and commit to it. We saw the benefits of such positive and constructive responsiveness (which we characterized as "capitalization") in Chapter 5. We strengthen interests and extend them over time by responding constructively to other people's positive stories of what is going right in their lives. These meaningful conversations are often the humble beginnings of lasting relationships. Curious people thus show a clear advantage in getting relationships started.

Very curious people also show a high tolerance for distress when trying new things, when uncertain of what is going to happen next, or when confronted with information that conflicts with their own beliefs. By being observant of what is going on inside them and around them, very curious people gain close proximity to their thoughts and feelings, and those of others, and this ability is also visible and highly regarded by others.

A tolerance for distress allows people to grow more comfortable with what are called "blended experiences"—those complex situations that simultaneously give rise to negative, stressful feelings and exciting, happy ones. A blended emotional experience might occur when meeting someone for the first time (the excitement of meeting a potential business partner with the anxiety of not knowing if they'll like you or your proposal) or when having a disagreement with a friend or family member where you feel angry but also relieved and positive about finally bringing a burning problem to light.

Not all conversations are pure joy. The curious, highly tolerant person has an advantage at negotiating emotionally challenging exchanges since they are able to remain open and engaged. When we aren't curious, or when our curiosity is overwhelmed by our anxiety or need for certainty, we are more likely to avoid or escape these situations. This ability to tolerate and work with negative emotions allows us to test and learn more about ourselves and creates opportunities to attract others to us. Likewise, being unable to handle teasing or an off day when someone is upset or angry is unattractive to others and limits opportunities to connect and deepen relationships.

Curious people extract a great deal of pleasure and meaning from socializing, and their social partners feel good in their presence and want to be around them. By recognizing these behaviors and trying hard to emulate them, all of us can become more engaged in our own social worlds and do a better job of triggering others' interest and attraction to us.

Nonverbal Displays of Curiosity Are More Powerful Than Verbal Ones

If you carefully attend to the reactions of others, you will notice that what people say is much less important than the nonverbal cues that "leak" out. I am reminded of an interesting, anxiety-provoking dinner in graduate school when students and their dates went to a restaurant with the faculty.

A professor was telling a raunchy story, standing with a wide stance, gesticulating wildly with his hands as he hovered over his sitting, captive audience. Now, some of these students were taken aback by his energy. In fact, one student was inching her chair away from the table little by little, trying to create some physical distance from this professor. But the professor clearly didn't notice, as he regularly stared in her direction to ensure she was listening. Thus began an entertaining dance between them—chair pushes back, professor moves in even closer, chair pushes back, professor advances again. Finally, the woman was practically bundled against the wall in a fetal position. Perhaps the professor enjoyed watching her squirm. More likely, he was so absorbed in entertaining the crowd that he failed to notice that although some of us were amused by the story itself, most of us were taken aback by his in-your-face, aggressive body language, and at least one woman was dreading every moment.

Two separate studies found that when people try to communicate a message, a measly 7% is conveyed by their words. The rest is determined how the words are spoken—facial expressions, posture, body movements, tone of voice, and so on. If you want to truly connect with someone, listening is insufficient; you must observe them. Don't expect people to verbally express their dismay at hearing another of your feeble stories without a plot or punch line. Politeness often prevails. If you really want to know what someone is thinking, pay attention and be receptive to the myriad of ways they are communicating with you.

When people aren't curious, it requires a great deal of effort and energy to try and get on the same page with them—what we might call high-maintenance interactions. There is persuasive research on the crucial skill of recognizing the emotions of other people and rapidly synchronizing with them as they change from one moment to the next. Paul Ekman, the distinguished emotion researcher, found that people who are open to new experiences and curious in general are more skilled at detecting and adjusting to these shifting emotions and thus are better at connecting with their conversational partners. Such synchronization allows people to work together and to understand what others are thinking, talking about, or doing. These are called low-maintenance interactions.

The beauty of this research is that it shows that detecting emotions is a skill that can be learned. The next time you are in a social interaction, try for two minutes to forget about the impression you are making. Forget about what you are planning to say next. Forget your agenda. Instead, be aware, be attentive, and be open to exactly what the other person is communicating. Is she tense, excited, touching you gently? If it's a romantic situation, is he mimicking your own movements and actions or teasing you in a playful manner—or at least trying his best? If so, you have an engaged, interested person. But look for changes. Do you see slight gestures to physically move away or make less eye contact?

For just these two minutes, get in sync with your partner's emotions and be respectful of what's being communicated. You might recognize how rare this practice actually is and how much you are missing out on in your conversations. Regular practice will make you a more engaged, interesting person to be around. If you don't believe me, see if the people who you find interesting and enjoyable in your life happen to be the most attuned to your shifting emotions, interests, and needs.

When we are actively engaged in our social world, we create and beget interest.

During my first therapy session with a high-powered lawyer, she mentioned being "burned enough times by useless conversations and first dates that started nowhere and ended nowhere" that she started giving people two minutes before ruling them out. This may sound callous, but every decision has an opportunity cost because there is something else we could be doing.

What I have found is that instead of passively waiting for disappointing conversations to end, we can actively transform them. In several studies, I brought two strangers into the lab to spend 45 minutes getting to know each other. To help them out, I gave them a deck of questions to take turns asking each other. Some couples were given questions that resemble the banal, boring small-talk that never seems to go anywhere at a cocktail party (How do you like the weather? What newspaper do you prefer to read and why? What did you eat the last time you went to the zoo?). Other couples, however, were given questions that required increasingly more self-disclosure to answer them, geared toward creating positive feelings and intimacy. (If you could only hold on to a single memory for eternity, what would it be? If a crystal ball could tell you the truth about yourself, your life, the future, or anything else, what would you want to know, and why?)

On average, everyone feels closer to each other after the high self-disclosure questions. However, we found that very curious people developed intense positive feelings and intimacy regardless of whether it was small-talk or talking about deep issues. Also, very curious people clearly wanted to explore a potential relationship because they were more likely to leave their phone number to see their partner again. Why? What happened?

In several ways, curious people transformed the small-talk and made it more interesting. First, they were attentive, open, and receptive to what their partners said and did. They resisted labeling, categorizing, and boxing in their partner based on first impressions. They were flexible in changing their view of their partner. Second, they encouraged their partner to share by discovering their interests and following up on them. Third, they wanted to keep the conversation interesting, and to do so, they were more likely to broach provocative and seemingly taboo topics (race, religion, politics, and sex) in a harmless, playful manner. They were less likely to censor themselves, sharing whatever interesting thoughts crossed their minds. These findings fit with the earlier studies I mentioned, which showed that very curious people are viewed by strangers and close friends as comfortable, animated, interesting, and fun.

I call this strategy "skipping the rungs of intimacy." Instead of slowly climbing the ladder from banter about the weather and what you studied in college in hopes of moving on to the playful, meaningful conversations everyone desires, why not jump right in? If you are interested, your partner is more likely to be interested (and if you are going through the motions, your partner likely feels the same way).

Because unwritten social norms are obliterated with this strategy, partners often get the butterflies as they wonder what is happening and what is coming next. Instead of viewing the anxiety and tension as discomfort, it is transformed into passion and excitement and a springboard for intimacy.

Not all the responses to this approach are positive, however. Some partners were uncomfortable and turned off. Obviously, if you are very worried about the possibility of being rejected, this high-risk, interest-enhancing strategy might not be the best fit. On the other hand, if this feels energizing, skipping the rungs of intimacy is a useful litmus test to see if someone is willing to play in your corner of the sandbox (with your unique interests, humor style, and all).

Whatever your inclinations, clearly expressing what interests you and what you don't want allows you to be authentic in your social world. You become the partner you want for yourself. Ultimately, being curious is about creating space to breathe and for positive feelings, interest, and meaning to emerge in conversations and relationships.

Communication Skills 101: Balancing Anxiety and Excitement

No matter how close we eventually become to someone, we always start off as strangers not knowing what the future holds. Before trying to trigger other peoples' interest, we need to ensure they are somewhat comfortable, we need to set each other at ease, and we need to sprinkle a few clues that direct our partner to our attentiveness, genuineness, trustworthiness, and interest.

When we "meet" someone online, we miss out on the ways we "read" and express our interest in another person, from facial expressions, body language, touch, and vocal tone, to the immediacy of attuning to emotional shifts of back-and-forth exchanges. Trust takes longer to develop online, and lying and deceit are harder to detect. As a result, abrupt rejection is quite common during that first face-to-face meeting. Intimacy developed online often has to be created anew in the real world.

Whatever your next encounter, here is a set of scientifically informed guidelines designed to harness curiosity for more meaningful connections and relationships. They focus on ways to break through the initial barrier of uncertainty to create common ground to get started. They can set the stage for you to shine and are likely to lead to some attention-grabbing moments where interest and excitement can spark and flourish.

Be Risk-Prone and Share

Anxiety is normal. Trying to make a good impression is normal. Trying not to look stupid is normal. So just talk. When people say they have nothing to talk about, that's not what they mean. They mean they think they have nothing interesting to talk about, and they are censoring the ideas that pop into their heads. Just let it out. Odds are your partner has the same concerns, so guess what, if you talk, you'll both be more comfortable. Now that it's clear you are going to do a number of stupid and foolish things, over and over again, go do them.

Get Your Partner to Feel Safe

Since nobody's motives are clear in the beginning, create a safe haven. Don't be overly critical, and avoid denigrating people from your past. Remember this mantra: The best predictor of future behavior is past behavior. Before you get to embarrassing, off-putting, or worrisome stories from your past, even if they are interesting, you want to make sure you've established a safe place to connect and get in sync with each other.

Discover Your Partner's Interests and Really Listen

Winning someone over is best achieved by getting them to talk about the things they care about the most: their interests and passions. Ask questions, and if there is common ground, emphasize it. Be constructive and focus on the positives of what they are interested in. When your partner opens up to you, share in kind what you are thinking, feeling, and what it makes you want to do. Support your partner's concerns, add and embellish with your own feelings and experiences, and check in to ensure the conversational adventure is two-sided.

Your Words Tell a Fraction of the Story

Be genuine in words and actions. Your posture, the sound of your voice, your facial expressions, eye contact, the distance between you and them, how and when you touch, all communicate emotions—positive, negative— and whether you are attentive and engaged.

Don't Be an Energy Vampire!

Show that you can absorb negative emotions and events and that you don't dish it out excessively or irresponsibly. Provide some reassurance that suffering is going to be minimal for the person talking to you. Disagreements, anxiety, and anger are all normal and par for the course during challenging first encounters. What matters is how you respond. Unless you go for the highly curious, skip-the-rungs-of-intimacy approach, start off with low-maintenance interactions that are full of energy and enthusiasm. Just as when you are dancing, you need to pay attention to your partner. If you fall out of step, observe their movements and fall back into place.

Be Open

Expect in advance that you might diverge in some major areas. If you view these differences as opportunities instead of dangerous roadblocks, you can get the best possible outcome in the situation. Don't avoid topics out of fear. So what if you differ in your views of race, politics, and religion; it might end up being a harmonious mix. Just stay open and actively listen to what they have to say. If you are afraid you might miss their point, check in and ask: "It sounds as if you are saying . . . is this right?" or "Let me see if I am following you . . . am I missing anything?" Everybody wants to be understood. Don't guess. Be open, and find out.

Focus on the Interaction, Not the Ending

Remember that you only shoulder 50% of a conversation. You can never control the outcome, only your actions and perceptions. Judge your performance, not the outcome. The worst-case scenario is you don't make a new friend or start a relationship. It might be a civil, friendly conversation where you both accept areas that cannot be bridged. It might end abruptly. I can strike up a conversation with a beautiful stranger while standing on line to get Rolling Stones tickets. I can be incredibly funny and ask all the right questions in all the right places. And yet, when I ask for a phone number, she shakes her head and turns away. I might think it's me and my bad haircut, or I might recognize that if she wanted a relationship, I executed my part perfectly, but she just wasn't interested. Cut yourself some slack. You are responsible for only half of what goes on. When you forget, you make your encounters needlessly more difficult.

The best thing you can remember is instead of trying hard to being an interesting person with interesting things to say, show the person you are with that you are genuinely interested in them. No promises or prior decisions need to be made. Interest and curiosity breed positive feelings that are contagious. Whatever the outcome and next steps in the encounter, when two people feel good, the door opens to the meaningful, interesting, revealing conversation that we all want.

The Search for Lasting Intimacy: Peas in a Pod or Opposites Attract?

As we've seen, making the initial leap to start a relationship, however casual, is challenging because we will always be uncertain about what other people think and whether they truly like us. That is why we prefer people who are similar to us in physical beauty, intelligence, interests, attitudes, and so on (the proverbial peas in a pod). It's an indication that we are acceptable and in sync with them. Rejecting a very similar person is like rejecting yourself.

When we focus on similarities, however, we overlook how little we ever know about anyone else. This factor of not knowing is, in fact, a key to strengthening all kinds of relationships. "Opposites attract" is a maxim that applies to many relationships. There is value in attending to what we don't know about people and how they differ from us.

Paying attention to differences requires us to adopt a new mindset; we break out of our assumptions and expectations to see the person or situation in a new light, bringing a fresh lens to the relationship.

So which is better—same or different? And how relevant is the answer, given the fact that in matters of the heart, what we "know" is not necessarily what we "do." Scientists know that regardless of what people say, they are not politically correct or rational when choosing new friends and lovers. Some people attract us, others repulse us, and we often strategically search for certain qualities that we may never have articulated.

If human beings desire growth and challenge (even if it leads to temporary discomfort), then similar people will sometimes be less attractive than people with unique knowledge and experiences that complement our own. In which case, we should be most attracted to new friends or lovers who offer maximum possibilities for expanding the self, rather than carbon copies of ourselves who already share our perspectives and interests.

Most of us experience a classic conflict when trying to form a relationship: We seek out the rewards of excitement and connection, and we want to prevent possible pain and rejection. When considering whether to attempt a new relationship, we seek the answers to two questions.

Are They Desirable?

We consider how attractive someone is in terms of how much we think we can expand or grow if we entered a relationship with them. There is certainly no reason to enter a relationship if we can get everything we want without them.

Does This Relationship Have a Chance?

We consider how realistic it is that we can form and maintain a relationship with someone and thus be able to expand.

In a study by Arthur Aron and myself, we created a fake computer dating service, but instead of romance, the goal was to help college students find friendship. All the subjects listed their interests, and we returned a week later to ask them to review a profile written by another person and judge whether they liked and wanted to meet them. Half of them were told that our ultra-reliable, matchmaker program determined that this new person was an ideal match for them. The other half weren't told anything. When people weren't given any information about whether a friendship was likely, they preferred people with interests just like theirs. But when they were told that a friendship was likely, they preferred people who complemented them with different interests. That is, when people were confident that a relationship was possible, they wanted to spend time with people who were unique, interesting, and who offered a chance for them to expand their horizons.

Other researchers came to similar conclusions. Married couples living together share a number of common goals that they work on together, from ensuring the fridge is filled to saving money for a new car. It might be useful when partners rely on different, complementary approaches to working toward goals and making decisions as opposed to being redundant with each other.

They make better progress toward important goals because they can divide the labor according to their particular strengths. If you and your partner plan a trip to Rome, it's more efficient if you can strategically divide the planning. Only one person is needed to find the best flights and hotels, someone needs to ensure that nothing is missing when planning and packing for the trip to avoid mishaps (checking whether passports are valid, ensuring enough sunscreen is packed), and someone needs to identify the "can't miss" adventures. Often we forget about the little strengths that make a trip, or any project, run smoothly. Do you realize how great it is to travel with someone who can jam clothes into every nook and cranny when packing suitcases so that fewer bags are needed? What about the wise person who reminds you to call your credit card company about the dates you're leaving the country? People who are good at preventing problems are invaluable.

Depending on the situation, being with someone who complements us is a blessing. Researchers found that married couples with complementary strategies for working toward goals and making decisions had the most committed, satisfying relationships. Partners with similar strategies were much less satisfied.

We can get a different perspective on this issue by looking at other situations where we want the greatest rewards and fulfillment. If we are putting together a dream team to complete a work project, start a new company, plan a perfect gathering of parents and kids, or win a ball game, what combination of people is going to work best? Researchers consistently find that for any task when at least two people divvy up the chores toward a common goal, people do better when someone has a style or perspective that complements their own. If I tend to be good with creative ideas (taking risks and willing to fail) and you tend to be good at seeing how all the details are organized and fit into the big picture (being responsible and avoiding mistakes), we will avoid turf wars and perform better.

Like an athletic team, each partner has a clearly defined role where they can make a unique, valued contribution to the team, resulting in greater personal fulfillment and more meaningful partnerships. To take advantage of the teams in our lives, we must take the time to uncover and incubate others' strengths and value these differences. Name them, and tell your partner what you see and appreciate. This isn't a single-shot exercise. We evolve continually, as do other people in our lives. Commit to continual quests to learn about people's strengths. Everyone has strengths. If you don't find any then look harder. You might be ignoring or dismissing what they do best and what energizes them. Assess their lives from their perspective, not yours. When you find their strengths, give them a name, share these insights, and refer to them regularly in conversations with them.

Three Ingredients for Lasting Love

Our scientific knowledge about relationships offers wisdom about choosing ideal lovers for lasting relationships. Of course, there are no money-back guarantees. In brief, there are three crucial ingredients:

1. Shared commonalities, which create a secure base (especially when core values are respected and appreciated)

2. Shared differences, which create opportunities for expansion and growth
3. Shared access to how each partner changes and evolves, which allows needs to be constantly acknowledged, adopted, and met

Be aware that there are areas where similarity tends to lead to successful relationships. Research shows that when it comes to physical attractiveness and intelligence, happy couples tend to be similar in these qualities. When it comes to cherished values or the guiding principles in a person's life, such as creating and caring for family and the centrality of religion, this is not the place to look for dissimilar partners. While it may make for exciting debates, this is the place where people are least likely to budge and thus, there are few opportunities to expand the self.

While it may be fun in the beginning, differences in these key areas often lead to animosity when big decisions loom in a shared life, such as how to raise children. The importance of knowing your core values cannot be overstated. It provides a foundation for making decisions, including who you might want as a romantic companion for the rest of your life.

Except for these fundamental values, finding partners who differ from you in several interests, skills, abilities, aspirations, perspectives, and resources *is* valuable. These differences and strengths give each person unique and valued roles in the relationship. These differences ignite curiosity and excitement, offering ample opportunity to explore, discover, and learn. Everyone wants companionship. Partners can spend time together, have fun, and grow in the process by ending up with two worlds to share and play in. Sharing unique worlds creates passion and commitment because the need for novelty and change is satisfied within the relationship itself.

Opposites might attract, but attraction doesn't mean that a relationship can be formed and maintained. Being opposite implies there is no shared space. Instead of thinking about people in black-and-white terms, it is better to think of similarity as a continuum. When we place people in categories, we overestimate how different they are and underestimate how well they might fit in our life space. Rigid thinking gets in the way of giving people a chance who might offer unexpected rewards. The ideal situation is two partners with complementary interests, strengths, and skills and each is respectful, wanting to understand what the other offers.

The Difficulties That Interfere with Strong Connections

Whether we relish having a huge social network or prefer a more intimate cast of a few close friends, we are hardwired social creatures who desire deep and meaningful social connections. Yet friendships or romances aren't easy to create, and they are just as hard to maintain. "Mishaps" and "failures" are difficult to avoid, but there are strategies we can use that can help us overcome them.

Loneliness

There is no shortage of people with unmet social needs, feeling as if they don't truly belong with anyone or any group. Loneliness, intense debilitating social anxiety, and fears of being abandoned affect nearly 25%.

As we get older and leave the overflowing social playgrounds of high school, college, and roommates, it becomes harder to meet our social needs. An even greater percentage of elderly adults feel lonely and alienated. Long-term friends die, the decline of mobility, and fewer calls and visits from kids and grandkids result in a great deal of solitary time. As if this wasn't enough, long-standing habits strangulate the willingness to try new things, so people have a greater reluctance to pursue new friendships as they age. Without initiating or being drawn into conversations with new people, relationships can't get off the ground.

Some people never enter into lasting, meaningful relationships because of fears of rejection, abandonment, intimacy, and other problems that make them feel undesirable. They may doubt their ability to make a good impression on and be welcomed into someone's fold. If unmet, that intense need for a sense of belonging lingers. These problems aren't limited to people trying to start relationships. It's common to see romantic relationships fall apart because people feel lonely.

Boredom

Topping the list of reasons for why couples decide to break up or seek therapy is that once blissful feelings fizzle out, they are replaced by a plague of shallow conversations and unbearable boredom.

The loss of self-expansion opportunities can lead to a sense of loss, and an itching sensation to find another source of intrigue and excitement. That nagging question starts to recur during quiet nights at home, "Is this it?"

Think about dating. First come wild nights of passionate sex, and in only a few years, the same two people are playing rock-paper-scissors to get out of scrubbing mold and mildew from the bathroom tiles. As routines set in, relationships lose their luster. Those initial positive feelings fade over time because our opportunity for self-expansion through our partner has reached a saturation point. We are confident that we have accessed everything we need to know about them, and the skills we wanted from them have been added to our tool boxes.

Instead of paying attention to the exact words and actions that "leak" out from our partner, we rely on our expertise about what we expect them to think, do, and feel. We rely more on the past as opposed to being in the present. Consider Gene.

Instead of telling his wife Samara about receiving a ticket for public indecency for changing out of his work clothes into his workout gear in a public parking lot, he kept it to himself. This is not because it didn't affect him. In fact, he spent much of the week wondering why Big Brother should care about his 10 seconds of nudity when that officer could have been off somewhere else preventing a grocery store robbery or pursuing a real criminal. Rather, he decided not to tell her because in his mind, he knew what she would say—that if he was more pleasant to police officers, they would let him off the hook; that if he handled the bills, he would know that we can't waste $200; and that he should stop acting like the college students he teaches. Thus, he kept it from his wife but shared the amusing story with other friends.

The problem is that when he does this, he ignores the impact of context and change. His wife tends to act in certain ways; that is, she has a stable personality. But this ignores how each situation is slightly different from any other, and even if she tends to be conservative and closed-minded *on average*, if he pays close attention, he'll notice that sometimes she is open and rebellious. Perhaps she would have found the story amusing, been aroused by the image of him naked in public, or been attracted to childish recklessness that's vastly different from his professor role.

By going on his hunch of what Samara would probably do, he may have missed out on a bonding opportunity. Withholding the story from her also deviated from his behavior during their courtship days. This slow, insidious move from exploring and discovering one another to making mindless assumptions and following scripted behavior (Gene is the irresponsible one, and Samara is the voice of reason) is not how relationships grow and flourish.

Researchers have found that when we suspect that the expansion phase of our relationship is fading (shuddering at the dullness on the horizon), the risk of infidelity increases. We start searching for other outlets to satisfy our needs. That coworker with beautiful blue eyes and perfectly carved calves isn't as attractive when an existing relationship meets our dueling needs for safety and growth.

As self-expansion opportunities start to decline, relationships suffer and beget the proverbial seven-year itch that often pains partners. Even seven years is overly optimistic, as recent longitudinal studies of couples show that many people start scratching after only two or three years in a romantic relationship.

This boredom isn't inevitable, though, if we continue to pay attention and bring a sense of the new into the familiar.

Mindless Scripts

Boredom affects other relationships, too. It's often what happens when we seem to "outgrow" friendships. You can think of relationships as an improvisational comedy troupe. When you have known each other for years, you don't feel the need to put in a lot of effort. It's easy to go on autopilot and fall back on scripted behaviors.

For instance, one friend might be the one who is always on the ball when it comes to ensuring long periods of time don't go by without getting in touch. The other friend might be the spontaneous, playful one who doesn't call or email regularly but is always ready to drop plans and begin an adventure with an hour's notice. This script falls apart when the person who is always making contact feels burned out and turns to other relationships where people care enough to reciprocate, or the extraverted, adventurous friend no longer feels the need to impress, grows tired of simply being "the life of the party," and begins to look for deeper connections.

By neglecting how much we don't know about someone as they navigate their own course through life, friendships wither away.

With a busy lifestyle, it's hard to nourish nonromantic relationships, whether with friends or family. Often in adulthood, we balance a number of separate social worlds that might include high school friends, college friends, work friends, and people we clicked with during random social encounters, who then turned into close friends. We need to consciously act to stay connected, we need to keep them updated on the ways we are changing and growing in the settings where they don't see us in, and we need to ask the same of them.

The Interest Gap

An alarming number of partners who claim undying love for each other are unable to hold on to those feelings for more than a few years. According to the U.S. Center for Disease Control, after 5 years together, 49% of romantic couples call it quits, and by 10 years, 62% of romantic couples living together separate or divorce. Of course, those couples who argue incessantly, resort to violence, and engage in infidelities should end their relationships, but they are in the distinct minority. Most romantic partners face other problems that are less obvious.

One of the main reasons for the stagnation, and sometimes the demise, of a relationship is a widening gap of shared interests and activities. Believing that relationships can rely on the past to function in the present is a recipe for failure. Offerings of novelty beckon us in every nook and cranny of our lives. When we experiment with new interests, our likes and dislikes broaden or strengthen (even if it is just by tiny fractions).

To preserve intimacy, we need to be in relationships where new knowledge and insights are sought, shared, and welcomed by each partner. Even if we want things to stay the same, we know that we are changing, so why should we expect anything less from people we care about. How we respond to change ends up being the legacy of our social world. Are we open and curious or closed off?

Working with Negativity

Many scientists and therapists who work with couples talk about the importance of good communication skills, problem-solving skills, and how to eliminate negative and increase positive thoughts and behaviors. If we want to avoid bad relationships, then we need to be honest and constructive when we talk to our partners and learn to handle arguments quickly and effectively before they fester and control the relationship. When it comes to spending time together with partners, there are studies showing that engaging in five positive behaviors (e.g., compliments or hugs) to each negative behavior (e.g., fighting or ignoring) is the ticket to satisfying relationships. Supposedly, if you fall below this magical 5:1 ratio, then you are pretty much on your way to a divorce, break-up, or endless inferno of suffering.

But once we leave the laboratory and enter the real world, relationships are a lot messier. There are relationships full of positive experiences and almost nothing to complain about (high positive, low negative) and emotionally intense relationships with the highest highs and lowest lows (breathtaking excitement but full of bitter, rage-infused arguments). Even more common, there might be nothing to pinpoint as negative or bad but in general, we just don't feel fulfilled (low positive, low negative). What this means is that we need to be clear about what we want because trying to eliminate the bad parts of a relationship is not the same as creating enjoyable and meaningful relationships. Some of those so-called negative behaviors might end up being good. In moderation, expressing negative feelings and getting them out in the open might sting, but it's just information to be digested and understood.

You might ask why that can't always be the case. You're right. If you or your partner suffer from emotional problems or express negative feelings too often or intensely, your relationship often suffers. Consider whether a therapist needs to be consulted. If emotional problems aren't part of the picture, research conducted by myself and others shows that withholding negative emotions leads to relationship problems. Without knowing what is going on in your partner's head, you lack insight into their concerns; intimacy is blocked and closeness is foreclosed. Expressing negative feelings is healthier and brings people closer.

Often anxiety, anger, and other so-called negative emotions are expressions of passion and excitement. With an open, curious mindset, we can learn from both the negative and the positive. Treating partners and relationships as an opportunity for discovery may be more important than any attempt to avoid negatives and load up on positives. In fact, I worry when people in relationships tell me they never fight and never get upset at their partner. This is a sign of holding back, concealing, hiding, and avoiding the working material that allows relationship to evolve. Work with instead of against the negative. This is something we will discuss in depth in the next chapter on the role of anxiety in leading a curious and fulfilling life.

If I could give you a choice between never having painful feelings but having no capacity to love or care, or being able to love and care but occasionally have painful feelings, which would you choose? Introducing novel activities and people into your life and that of your partner is sure to be a blend of emotions with interest and excitement mixed with anxiety, nervousness, anger, drowsiness, and guilt (depending on that mental chatter that no one else is privy to). Being open means being willing to experience and absorb emotions, positive and the negative, because they are stepping stones to discovering, understanding, and growing in relationships.

I am convinced that we don't have to banish people who fail to immediately capture our interest. We also don't need to bail out of strong relationships when they lose their luster. We have to be willing to accept a certain level of negativity from other people and in ourselves. So what can we do to transform these unwanted negative thoughts and feelings so our relationships are more interesting and rewarding?

The Magic of *Adding* Relationships and Interests to Sustain and Energize Healthy Relationships

Just as we can't leave babies to fend for themselves when we escape for a luxurious spa weekend, and teenagers will shut us out if we don't stay up-to-date on who they are becoming, our own relationships need to be regularly checked on and fed.

There is substantial evidence that people who spend the most face-to-face time with their partners experience the most satisfying relationships. But it is not just about the amount of time spent together, it is how that time is spent. Is it spent rigidly following well-grooved habits and routines? Or is it spent in new and novel situations and attending to variations in interests and emotions in one another, just as we attend to growing children?

Pleasant experiences such as eating out and going for walks offer short-term mood boosts but are often easily forgotten. It is the active attention that engages our curiosity and openness that counters the decline in passion and relationship satisfaction. This attention is a hallmark of couples who remain invigorated long after the honeymoon period is over.

Freddy is a retired 82-year-old who has been married to Betty for 54 years. On the surface, he appears fragile. Several seconds pass before his shaking legs pull up behind his cane. Once he began to speak, I quickly realized this first impression was deceiving. He retired because he "got fed up with work." Freddy's eyes bulge with pride as he mentions that he is president of a local comedy club, took a part-time job as a pharmacist, works out in the gym, and enjoys beating the stock market with risky investments.

After the interview ended, he couldn't resist telling me about an up-coming weekend cruise with Betty through the Panama Canal "because the lock system is utterly fascinating" and he wants to "see it in action." Asked about his intense energy, he says "you can have and do anything when you have a spouse to go with you . . . you need to feel like you are loved and love them back." Freddy shows it's possible to sustain high levels of excitement in a relationship long after the early phase of getting acquainted is over. For Freddy, nearly everything he does involves his wife or is shared with her. Freddy isn't just going for walks in the supermarket with her. He does things he is passionate about and, in some capacity, Betty always joins him for the ride.

If we look back at the early phases of relationships, our hearts and minds were pulsating with adrenaline and passionate excitement like a dryer with a pair of shoes bouncing around in it. In the beginning, we grew because our partner brought novelty, challenge, and variety to our lives—and we did the same for them. Using the same logic, as self-expansion opportunities fade, it is imperative to find new sources of interest and excitement.

If we are doing something novel such as designing a bird sanctuary in the backyard, test driving a Lamborghini Murcielago LP–640 (one bad-ass motorcycle!), picnicking outdoors before a summer concert, or tasting wild-boar risotto, we add an entry to the encyclopedia holding all of our personal adventures. Activities can range from the exotic to the ordinary, as long as we attune to the novelty and twists that a familiar situation has to offer (taking on the "beginner's mind"). We increase the amount of information and knowledge that we had before and stretch our skills when we do something beyond the confines of what we already know.

In the comfort and safety of our romantic partner, the anxiety and tension we experience in new and risky situations often makes us feel alive and excited. It triggers the feelings we had during our very first dates when we had no idea what the future held and if everything would turn out okay. The highly desirable state of expanding and being fully alive is kindled afresh and becomes linked to our partner and the relationship.

This entire process of connecting positive experiences to our partner and the relationship happens outside of our conscious awareness. That is, we often have no idea why we feel a surge of loving feelings toward our partner after doing something interesting together.

To test these ideas, researchers assigned relationship-enhancing exercises to 53 married couples. Any change would be profound because couples had been together for an average of 15 years! First, the participants looked through a long list of activities, rating them for how pleasant or exciting they would be to do as a couple. What someone rates as exciting is not always pleasant because when you try something new (out of your comfort zone), often you feel unwanted anxiety and are afraid of looking foolish or failing. Activities typically rated as very exciting were dancing, going to concerts, meeting new people, and thrill-seeking sports such as jet skiing; activities typically rated as very pleasant (but not exciting) were visiting friends, seeing a movie, and eating out.

Couples were randomly assigned to spend 1.5 hours each week for the next 10 weeks taking part in "exciting" or "pleasant" activities based on their own ratings. It was important for couples to choose from their own list because what's exciting to you might be dreary and dull to me. Other couples were in the "no-intervention" group and told to do what they normally do for 10 weeks.

What these scientists found is that simply spending time together is not enough to enhance the quality of a relationship and that doing exciting things together is more valuable than doing pleasant and relaxing things together. Couples doing more exciting activities experienced a greater increase in marital satisfaction over the 10 weeks compared with couples who injected more pleasant events into their lives or those who stuck with their usual routines. The findings are profound because after only 10 weeks, couples viewed their 10-, 20-, and even 30-year relationships differently, and when excitement was added to the mix, they found them to be more satisfying.

To determine exactly why taking part in exciting activities was linked to greater relationship satisfaction, researchers asked 112 couples who had been together for an average of 6 years to answer questions about their relationships such as "How bored are you in your relationship?" and "How exciting are the things you do in your relationship?" Couples doing a greater number of exciting activities viewed their relationship as more satisfying. Their exciting activities disrupted the day-to-day monotony that can creep into any relationship.

As a rigorous test of whether novel and exciting activities can demolish relationship boredom, we can look at what happened when the research couples were sequestered in a laboratory to study the role of these new activities. With this approach, researchers controlled the situation to make sure that all the couples engaged in the same novel, exciting activity (to rule out the idea that some couples just do cooler things).

One series of studies involved dozens of couples who were together for at least one year. Some couples took part in a seven-minute novel, shared task. With their wrists and ankles tied together, they were instructed to drag their tangled web of limbs from one side of a soft, foamy room to the other side. A huge pyramid of rolled yoga mats was piled across the middle of the room, and their only option was to climb over it. As if this wasn't hard enough, without using their hands, they had to carry a foam cylinder with them to the other side of the room.

At the end of the task, couples said it was extremely bizarre. Of course, you would be leading a pretty unusual life if you claimed otherwise. The other group of couples spent seven minutes taking turns crawling to the pyramid of mats and returning, back and forth, until time ran out. They did the task together as a couple but without the added excitement of being tied together, having to climb over the pyramid, or carry the cylinder. Basically, they took turns crawling on a mat (how exciting!).

While we might expect couples doing the novel task to feel better in the moment, an interesting thing happened. They didn't just feel good. They actually viewed their relationship differently than couples doing the mundane task. Couples viewed their long-standing relationships as more satisfying after a mere seven minutes of doing a unique activity together. The shared positive feelings carried over to how the partners viewed their entire relationship, seeing it as more rewarding, satisfying, and meaningful.

No such relationship improvements were found for those doing the dull activity together. In fact, many of the couples viewed their relationship as *less satisfying* after the dull activity. That's important. If you fail to fill your free time with interesting activities, boredom and dullness spill over into how you judge your partner and relationship. For this reason, I encourage you to re-read Chapter 4 to incorporate techniques for triggering interest in the moments of your life.

What this new science of curiosity can teach you is that these growth opportunities are available for the taking. Focus on them, share the experiences with others, and make sure they stick in your memory (activate that hippocampus!). These growth opportunities do more than keep relationships intact; they keep them energetic and alive.

Another group of therapists working with couples realized that waiting for relationships to fail was a failed approach to enhancing marriages and intimate relationships. They started working with healthy couples who had no psychological problems to teach them the techniques of mindfulness therapy. Without going into too much detail, mindfulness therapies teach couples to be open and curious about their thoughts and feelings, to work with negative feelings (instead of trying to avoid or eliminate them), and to live in a way aligned with core values (see Chapter 5). Other skills include cultivating self-compassion and compassion for others, and learning yoga and touching exercises to bring couples into calm, loving synchronicity.

Several clinical trials were conducted to see if mindfulness therapy for couples worked, and so far, the verdict is good. The couples studied were more satisfied with their relationships. As to why it worked, the intervention pushed couples to participate in novel and interesting things together and appreciate qualities about their partners and their relationships that they had previously *overlooked*. For a relationship to thrive, we need to ensure that flexibility is built into our scripts. We need to be continually interested in peeling back and penetrating the surface of our partner, and when we do, we relate to them mindfully.

Feeding Intimate Relationships by Bringing Others into the Mix

No, I am not talking about being a swinger and swapping partners in a risqué Vegas night club. I am talking about the effects of friendships with other couples on the interest and closeness in your own relationship. As if dating wasn't hard enough, becoming friends with another couple involves my liking both people, my partner liking both people, and each of the members of the other couple liking my partner and me. For a foursome, there are a staggering number of missteps that can lead to failure. To create a working foursome, especially if there is only one hole (e.g., my partner doesn't like one of them), consider the strategies discussed in the section of this chapter on triggering interest.

Once you find another couple to spend time with, besides adding people to your social world, you create an opportunity to enhance your own relationship. Here's why.

No matter how much time we spend with someone, we are privy only to pieces of them and what they are capable of. Our behavior influences and changes the person we are with and thus

- We can never know exactly who they are (this is called the "observer effect")
- We regularly overemphasize what we expect to see
- We regularly fail to recognize and underemphasize what we don't expect to see

As an active member of the relationship, we lose our objectivity, failing to see our partners as they are (as other people who are less familiar with them would see them). If we want to re-create that initial rush of positive feelings and rapid expansion during the early days of our romance, if we want to be interested and our partner to feel the same, we have to change this mindset—from mindless drone to curious explorer.

I can learn a great deal about my partner by watching her in action with another couple. Hearing my partner tell a story that I heard before can take on new meaning when I watch other people listen intensely, laugh hysterically, or be enthralled by my partner. Their attraction to my partner intensifies my own. As social creatures, we value the opinions of people we care about. Although it sounds crude, my partner's stock value goes up as I appreciate anew what is so amazing about this person whom I decided to bring into my life.

Besides becoming increasingly grateful and attracted to my partner, I can discover new things. My partner is going to learn about my likes and dislikes and because of that, she is not going to share everything about her past and everyday actions. We all learn what gets a positive reaction, what doesn't, and then modify what we do next based on these rewards and punishments (not responding to what my partner cares about is essentially punishment by neglect).

When we spend time with another couple, two new people with unique tastes that differ from my own are going to draw out stories from my partner that I might have never heard before. As I witness this, my partner becomes novel and interesting, and this effect lasts well beyond any conversation or situation. And it's not just the telling of stories—new people affect how my partner is going to think, feel, and act.

With another couple, my partner might be more assertive and dominant (as opposed to falling into a quiet, more submissive role with me) or goofy and playful (as opposed to tending toward being serious and intellectual with me because those are the kinds of conversations that I regularly draw her into). We can look at our partners from new perspectives, collect new details, and begin to grasp how complex they really are compared to the false idea that we already know everything about them. If you are willing to be open and curious, you can enjoy realizing that there are things you think you understood about them that you didn't, and there is always going to be unknown information awaiting discovery.

Richard Slater, a psychologist at Stanford University, extended research about the ways two people can refuel their interest and excitement after being together. He matched 60 unacquainted married couples into pairs, and had foursomes do the 45-minute closeness induction procedure where they picked up cards with intimate questions on them and took turns answering them. Other couples worked their way through 45 minutes of mind-numbing small-talk questions. His results were amazing.

First, couples generally felt a great deal closer to other couples after the closeness questions than those who worked with the small-talk questions. In fact, many of them appeared to be on their way to forming real friendships, passing along phone numbers to meet after the experiment was over. In contrast, everybody in the small-talk group appeared to be ready to depart quickly when the session ended; they weren't happy or upset, they just felt a lack of energy throughout the conversation.

Second, each couple in the closeness group felt better about *their own relationship* after the interaction. They felt more satisfied about the partner they came in with after sharing intimacies with another couple.

Third, the reason they felt more satisfied about their relationship was that they felt a surge of positive feelings, they found the entire task to be novel and interesting, and they "learned new things about their partners." In a mere 45 minutes in a sterile laboratory, talking to another couple led them to grow (even if only a tiny bit) in their relationships. In the immediate aftermath, they reported viewing both their overall relationships and their partners as more interesting and satisfying. Although we don't know how long these thoughts and feelings last, just the fact that we change long-standing views suggests these new experiences are portals to seeing our partner from new angles and perspectives.

As my mentor and research collaborator Art Aron said to me, "I think one of the ways that couples can make their life richer, fuller, and cope with boredom, is with other couples. It is certainly true for me and Elaine. We have one couple we are really good friends with and they bring us to things we would never have done, and we do the same for them."

The research and examples I have been focusing on look at couples, but your relationship can expand by bringing any other person or persons into the picture. But be sure to look for curious explorers. Look for interesting people, not someone who is going to put you in a foul mood.

Sometimes the worst people to get us to do new things or give us advice are the people who know us best, our romantic partners, best friends, or parents. As soon as someone else says the same exact thing, we listen and are motivated to go do something new. When we are very close to someone and see them as an extension or growth of ourselves, we often require another person who is more removed and objective to ignite the flame. Other people bring the new into our relationship. We get to peak through the beginner's mind. This is one of the simple truths for preserving novelty and interest for the long haul.

Just remember, you need to share new people in your life *with* your partner for it to influence the relationship. An early death to a relationship is almost assured when there is a widening gap of unshared interests, people, and lives.

If you turn to the appendix of this book, you will find detailed instructions for applying these scientifically informed strategies to invigorate your own romantic relationship.

Final Thoughts

What this research says is that being more adept at appreciating the mysteries and uncertainties of relationships and life is where true fulfillment can be found. Go beyond the need to predict, understand, and control your social world. Attend to what you don't know, expand the boundaries of who you are and what you do, follow your instinct of what is interesting to you and what is interesting to other people, and this will lead to positive changes in your social interactions and relationships. Understanding these concepts may be all you need.

When it comes to romantic relationships, some couples might need a therapist to guide them. Note, however, that the research findings and insights in this chapter are very different from what many couples want from therapy. Many couples want answers. They want to reach a sense of certainty and closure. They want to be able to predict their partners' every move because it releases them from feeling anxious and uncomfortable. They want to be absolutely certain they are going to stay together and work it out. That might work if you are interested in staying together and couldn't care less about the quality of your time together.

My guess is that most people want more than physical proximity to others. We want to make exciting and meaningful connections. We want our romantic relationships not only to last but to continue to contribute to our ongoing growth, happiness, and well-being over time. The path offered by science points us in the right direction.

7

The Anxious Mind and the Curious Spirit

Twenty years from now you will be more disappointed by the things that you didn't do than by the ones you did do. So throw off the bowlines. Sail away from the safe harbor. Catch the trade winds in your sails. Explore. Dream. Discover.

—Mark Twain

Do not be too timid and squeamish about your actions. All life is an experiment. The more experiments you make the better. What if they are a little coarse and you may get your coat soiled or torn? What if you do fail, and get fairly rolled in the dirt once or twice? Up again, you shall never be so afraid of a tumble.

—Ralph Waldo Emerson

We possess an old body crafted by natural selection to do more than just explore and discover—it's also designed to help us survive. Each time we recognize something new and are uncertain about what might happen, we often feel a combination of curiosity and anxiety.

We regularly find ourselves in battle—with ourselves. We get signals from the anxiety system, and our bodies gear up for danger, preparing us to fight or escape. If we avoid the situation, then nobody gets hurt. At the same time, anything new is a potential treasure trove. If new things hold the possibility of hidden treasure, then how can we walk away without exploring and sucking the nectar dry? And so we duel it out between avoiding (playing it safe) and exploring (seeking out rewards).

Everyone has moments when they are fearful, nervous, and worried that bad things are going to happen. Depending on the intensity and how you handle it, anxiety can be beneficial or it can be destructive. Learning to work *with* anxiety makes all the difference in creating a high concentration of fulfilling moments wrapped inside a fulfilling life.

One of my students, a grandmother named Christine, shared a story that encapsulates life into the moment when anxiety and curiosity collide.

One day I took Brooke, my 8-year-old granddaughter, to the toy store. Brooke was interested in many things as we walked the aisles and seemed fascinated by the long row of dolls. She happily pointed to each one and announced, "Baby!" with a big smile. Suddenly, we both sensed movement behind us.

We turned around to see a young couple walking with a small boy, who looked to be about Brooke's age. The couple and the boy stopped when they saw us in their path and Brooke smiled at him. The boy's parents stood quietly, probably thinking that the two kids would have a sweet encounter. Instead, the boy rushed full bore ahead, stretched out his arms, and shoved Brooke as hard as he could. Brooke went flying backwards, sliding on the tile floor until her head hit the shopping cart, where she came to a stop. I quickly picked her up, cradling her head and talking softly to her as she screamed.

The parents apologized, and I could see they were appalled. I also heard them tell the boy that they were not going to buy a toy for him that day because of his behavior. They left and I stood soothing Brooke.

The young parents returned shortly after the episode. They

wanted their son to make amends. I gave them credit for trying to teach their son to do the right thing. The parents made sure their son was not positioned to charge Brooke again and encouraged him to apologize. Brooke stopped crying, but her face was wet with tears.

What struck me was this: When the boy returned, Brooke didn't smile or look at him expectantly. The openness that had been there at their first meeting was gone, replaced by an expression of wariness and a tighter hold on me. Brooke signaled to me that she was unwilling to get down as long as the boy was present.

When Christine told me this story, I recognized that all of us are confronted with these conflicts, when events impinge upon our natural desire to seek out rewards. I wondered how often in life we are open and curious, only to be met with disappointment or a need to be wary. Because we don't want to experience anxiety or fear, we shut the door on possibilities for future experiences.

Being overruled by anxiety from time to time is not going to ruin your life. In fact, it's often the right decision, as it was for young Brooke. However, patterns can develop over time, and this is what we want to examine. Do we take calculated risks and aim for the biggest possible rewards? Or do we go out of our way to make sure pain, errors, and failures are kept to a minimum?

If anxiety overwhelms our curiosity system, dictating what we do, then our plans for the future get hijacked. On the other hand, if we feel anxious, recognize that the future is uncertain, and willingly take part in the things that naturally interest us, a different life narrative emerges. This chapter is about understanding and harnessing curiosity and anxiety. By doing so, you can unleash your maximum potential, ensuring that a greater proportion of your energy is spent doing things that offer satisfaction and meaning.

Curiosity as an Anxiety Antidote

Beth Comstock, President of Integrated Media at NBC Universal, attrib-
utes her career accomplishments to "her passion for new challenges." She
explained that it's easy to get anxious and shy when you are surrounded
by intelligent people and making decisions where millions of dollars are
at stake. Asked how she copes, she said, "I found that curiosity propels me
forward. Sometimes I look back on a situation where I was reluctant to
speak up, even when I knew I could have added value, or was afraid to meet
someone new and regret the missed opportunity. That's what I tell myself
now: 'You don't want to miss this opportunity. Get in there.'"

Just like Beth, curious people show a readiness to plunge into experiences
regardless of how it feels and what it looks like. It's not that curious people
are free from anxiety. They get anxious just like everyone else (anxiety
can range from most of the time to rarely). Instead of focusing on how
much anxiety they feel, curious people benefit from what they do with
their anxiety. Curious people act on their curiosity and explore whatever
is intriguing them, taking those anxious feelings and any discomfort they
bring along for the ride. Instead of relying on premature descriptions of
whether feelings are good or bad, people are good or bad, or situations are
good or bad, curious people test things out. They show a willingness to
explore.

By enhancing curiosity, we can become more comfortable tackling
change and difficult life circumstances. For instance, young children are
often scared to go to the hospital—and for good reason. Hospitals are
filled with the walking wounded, moaning and bleeding in the hallways;
nurses carrying needles the length of a baby's torso; and rooms filled
with beeping sounds, pulsating lights, and strange-looking gadgets. Two
researchers in Australia wondered if they could reduce children's hospital
fears by giving them a chance to physically play with a variety of medical
equipment and ask questions about how they worked.

The author's hypothesis was simple: Children who get to be curious
and explore scary-looking medical equipment on a safe visit to the hospi-

tal might be less anxious when they need to be treated for a medical condition. A group of kids aged 3 to 5 played with chemotherapy glasses, intravenous bags, syringes without needles, blood collection tubes, and so on. The children asked questions and were exposed to as much information as they desired. Some kids had prior experience in a medical setting and showed off their knowledge to the others, "Let's try getting water out of the IV drip . . . there, the medicine is coming out of it." Others used the medical equipment to play games—the syringe was a crowd favorite to water the plants or to spray each other. The authors found that more curious children (those who asked questions and manipulated objects) learned more about the equipment and procedures and made more positive comments about doctors and hospitals. It was clear that children "have a naïve, joyous curiosity about medical equipment." The more kids are allowed and urged to explore and play, the more misconceptions about the dangers of intensive medical equipment and hospitals can be replaced with accurate information. Thus, children gain insight and cope better when they require hospital services.

Thriving with Strengths and Anxiety: Tuning Our Exploration and Anxiety Modes

Some aspects of our lives might need to be changed. We might, for example, need to eliminate overly aggressive, oppressive, violent, or toxic people from our social world. Other times, there may be nothing we can do to bring about change. When we are unable to differentiate acceptance and change, we suffer. This delicate balance is captured by my colleague, Hank Robb, who slightly reworked the widely known serenity prayer:

> Let me seek acceptance of life as I find it, even though I may not approve of what I find, wisdom to see what would be good to change, willingness to act as well as willingness to follow through, and gratitude for the opportunity to try to live my life as best I can.

Consider spending time with people, being in places, and doing things that are interesting, enjoyable, and personally valued even when there is a lot of anxiety. In fact, we can be at our best and use our strengths while having anxiety.

I am going to ask you to think about the other options available to you when you struggle with unwanted, unpleasant feelings such as anxiety. Dr. Steven Hayes created a mindfulness-and-acceptance–based intervention called "Acceptance and Commitment Therapy," or ACT, to help people become psychologically flexible and commit to behaviors that are aligned with their deepest values. Controlled studies comparing ACT to cognitive therapy and medication found impressive gains in treating problems as wide ranging as anxiety, pain, smoking, diabetes management, adjustment to cancer, and epilepsy. Regardless of whether you are suffering from psychosis or seek to cope better with stress and anxiety, it's helpful for people to begin by focusing on a metaphor he calls "Two Systems"—the curiosity and the anxiety systems.

For just a moment, think about the situation where you feel the most anxious and uncomfortable. Maybe it's when you're speaking in public, going on a job interview, talking to an attractive person you want to date, or thinking about how little money you have at your disposal. Make the situation personal. Everyone has something that causes them at least a little anxiety.

Think about what you most want to do when you feel that anxiety. Do you want to run? Do you freeze? Do you want to hide? Do you want to conquer the moment? With this personal image in mind, consider the following metaphor:

Think of the anxiety and curiosity systems as two knobs, just like the controls for volume and balance on a stereo. The one most familiar to us is the Anxiety knob. In theory, this knob can go from 0 to 10, but you might be saying to yourself, "I can't seem to get it lower than 5, and I really can't stand it when it's 8 or higher. It's uncomfortable. I've got to do something about it." This is probably the knob you spend most of your time and energy on.

There is another knob hidden in the back that's harder to see. It's called the Explore knob. Just like the Anxiety knob, it can go from 0 to 10. It might be the more important of the two because it's the only one that you can control. It refers to how open you are to accepting your own experience as it is. When Anxiety is at a 10 and you're trying to make it go down, to make it go away, then you're unwilling to feel the anxiety that's registering on the dial. Looked at another way, you've set the Explore knob to 0.

This is a terrible combination. If Anxiety is high and Explore is low, there is nothing you can do to change your situation. Anxiety can't go down. You simply can't move it at will. You can't just tell yourself to be calm. In fact, when someone says this to you, it usually just makes you more anxious. When you try to relax to feel less anxious, it can make you more anxious because you start wondering why you can't calm down.

You've been trying to play with the Anxiety knob for some time now. It's not that you're not clever, it simply doesn't work, and it can be frustrating to keep at it. So, keeping your anxiety-producing moment firmly in mind, I now want you to shift your attention to the Explore knob. Unlike the Anxiety knob, which you can't move at will, the Explore knob is something you can set anywhere. It's not a feeling or a thought; it's a choice. You may have kept it set at low because you didn't know you could move it, or never thought to. But you can.

What you need to do is set it high. Now imagine your highly anxious situation, still on high, but now your Explore knob is turned all the way up. What do you notice now?

I can tell you exactly what will happen in real life, and I will swear to it. If you stop trying to control Anxiety, your Anxiety will be low . . . or it will be high. And when it's low it will be low, until it's not low, and then it will be high. And when it's high it will be high, until it isn't high anymore. Then it will be low again. I'm not teasing you. There just aren't good words for what it's like to have the Explore knob set high. I can say one thing for sure, though—if you want to know for certain where the Anxiety knob will be, there is something you can do. Just set Explore very, very low, and sooner or later when Anxiety starts up, it will lock into place and you will have plenty of Anxiety. It will be very predictable.

If you move the Explore knob up, Anxiety is free to move. Sometimes it will be low, and sometimes it will be high, and in either case you will have the awareness you need to take action to move in the direction of your values and what you aspire to do with your life.

This is a special kind of mindset that falls under the notion of mindfulness (which we explore in Chapter 2) that offers profound benefits.

I met a biologist who was fascinated with frogs but scared of the dark and enclosed spaces. She was in a quandary because there is a certain cave where an endangered frog species was recently found. The opportunity to answer a host of questions excited her. Her excitement was so strong that she pushed herself to make the trip into the cold, dark cave. She found the frogs, published an important paper that drew attention to their plight, and—while she was underground—found a gorgeous waterfall. She cherishes the entire adventure and tells anyone who will listen.

It's an important story that any of us can connect with in some capacity. Anxiety and desire are two sides of the same coin. Forgo anxiety and you will be unable to tackle hurdles that get in the way of pursuing goals aligned with your values. Don't wait for anxiety to go away because it will never happen.

Anxiety as a Fear of Making Mistakes

One reason that we don't get into the fray and explore is because we're anxious about making mistakes (some of us get more upset than others). We get so focused on not looking like idiots that we get trapped inside our heads talking to ourselves. This leaves us with less energy to be an active part of whatever we are doing. Worrying about making mistakes prevents us from being interested and creative.

There is a nifty study by Ellen Langer, who asked Harvard college students to give unprepared speeches to an audience. She wanted to see whether being open and curious could transform public speaking anxiety. How? By asking speakers to change their mindset of what constitutes a mistake. Langer randomly assigned students to one of three conditions: (1) the "mistakes are bad" condition, where they were told not to make a mistake; (2) the "forgiveness" condition, where they were reassured that mistakes are fine and instructed to purposely make a mistake; and (3) the "openness to novelty" condition, where they were told to incorporate any mistakes they made into the speech itself and instructed to purposely make a mistake.

Speakers in all three conditions gave a talk in front of a room full of people and were told that they would be judged on how well they performed. The results? Speakers in the openness-to-novelty condition judged themselves as more comfortable and rated their performance better than speakers in the other conditions. Moreover, the audience also judged the speakers in the openness-to-novelty condition as being more composed, effective, creative, and intelligent.

Knowing this, perhaps we can cut ourselves some slack. Confucius said, "being ashamed of our mistakes turns them into crimes." Allowing ourselves space to make mistakes and being open to whatever direction the present takes permits us to be better speakers and conversation partners.

I think there is another reason why mistakes and errors draw people in: People find us endearing and attractive when we make a few mistakes. We become more present, alive, and human, and less a preplanned, canned machine.

Odd, unpredictable happenings trigger interest, pulling us from our slumber into the present moment. This is why people are willing to pay exorbitant prices to hear their favorite bands live. We can buy ten CDs of our favorite band or pay for one live performance. Why would we want to see them live on stage? After all, those CDs are edited to be absolutely perfect so you get to hear the best possible performance. We pay to see bands perform live for the spontaneity, the possibility that anything can happen.

The same goes for speeches and conversations. We're attracted to real, live, flawed, creative people. When people are open to experimenting and playing with so-called mistakes and errors, they enjoy it more and so does their audience. This is why crowds hollered in excitement when comic actors like the regulars on *Saturday Night Live*, Jon Stewart of *The Daily Show*, and the late Harvey Korman of the *Carol Burnett Show* would get caught in the moment and laugh at their own jokes. I suspect one reason these shows are so beloved is because the audience always has the possibility of stumbling on novelty. A genuine relationship develops, and it keeps people glued to the television.

Curiosity and Our Social Roles and Identity

Anxiety, when it isn't balanced with curiosity, intensifies the need for closure and certainty, which plays out in every sense of who we are (our identities) and how we connect with others (our social roles). When we lack safety, when we're worried about our acceptance in a group, when we fear making mistakes, it can seem easier to simplify our identity.

We may define ourselves as mom, homemaker, gay, Latino, or war veteran. We may label ourselves as reserved, caring, or responsible. Far too many people experience great strains from the conflict of managing dual roles such as work and family, or surviving the inner city while trying to complete high school and go on to college. Only by seeing the various angles of our identities and how they work together can we successfully search for meaningful ways to spend our time and energy.

I had an opportunity to talk with a man struggling with his conflicting identities. Rasheed was a gay man tucked into a compact 5'5" frame. Although he is very religious, it's been hard for him to reconcile his faith with the fact that gays are often criticized and demonized by the church. He said, "nobody in the church knows about me and I figure it's best to stick with a don't ask, don't tell policy."

At the gym, he started working with a personal trainer to change the shape of his sphere-shaped torso. In only three months, he lost 20 pounds. As a motivational tool, he carries around a picture of a shirtless male model with the physique he is after. Proud of this "future self," he wanted to show me the picture before gasping, "Oh man, I left the picture in my church bag! I should probably take that out before someone sees it. It's so hard to keep my life wrapped up." I remember feeling saddened that his motivation, his big accomplishment of battling his weight, has to be hidden from people who respect and care for him at church for fear that they might learn of his sexual orientation. Playing hide-and-go-seek with his identity, Rasheed is constantly at battle with himself.

I imagine scenarios where Rasheed adjusted the settings of the Anxiety and Explore knobs differently. Despite being anxious about what people at his church might think if they found out he was gay, he carries around that picture of his ideal body, motivating him to be a better version of himself. I suspect this pursuit of a healthier body makes him a better person, whether at church, at work, or with his friends and family. I hope that he experiments with being authentic. If they see the picture, they might not assume he's gay. They might see the motivated, brave person I see. If they do discover that he is gay, they might welcome him with open arms. If they don't, then this little experiment might give him a good reason to search elsewhere for social support, morality, and spirituality.

Instead of constricting his life space, these alternatives open up space for him to move in the direction of his values without ever intentionally trying to move the Anxiety dial. He might end up feeling less anxious, but this is secondary to working toward important goals. The best way to get on the road to a fulfilling life is to raise the Explore knob, not the Anxiety knob.

Seeking safety and security might reduce anxiety and pain in the short term, but it is nonsense to believe that we must remove or manage unpleasant experiences before we can attempt to invest in what truly interests us. We can begin as soon as we are open and receptive to what we want our lives to be about and accept that difficult thoughts, feelings, and memories are part of the journey of moving toward something that matters.

Curiosity Helps Us Work through the Conflicts and Uncertainties in Our Relationships

Don't take my word for it, the research shows that being curious can help resolve our conflicts so that we can find meaning, maturity, and other positive elements from difficult events. Exploring our inner world is a backdoor route to make sense of difficult personal material as we try to return to a state of emotional equilibrium.

Returning to Aron's work on romantic couples from the last chapter, in one study, he wanted to know whether participating in a novel activity improved handling of negative emotions. You might recall that couples worked their way across a gymnasium floor while carrying a foam cylinder without using their hands. Before and after the task, they took part in an emotionally charged conversation—plan a vacation together or discuss what you would do with $15,000 to renovate your house. Some difficult emotions surfaced, yet after the interesting task, the couples showed less hostility and less anger in their exchanges. Instead of erecting barriers when difficult emotions arose, couples in a curious state of mind just rolled with it, working with them.

In the context of intimate relationships, research shows that curious people are good friends and partners. They don't get overly anxious or feel threatened by what happened in the past or what might unfold in the future because they see this as a mystery to uncover. Curious people don't get overly anxious or threatened when they don't know how things are going to unfold. In turn, partners feel respected when someone is willing to trust them even if absolute assurances can't be given (because no one knows what the future holds).

When people need certainty, they have a hard time coping with their anxiety because they cannot fully and completely trust someone. They view relationship partners in black-and-white terms: completely trusted or a louse. It's tough to be the partner of someone with these extreme beliefs. In a similar vein, they prefer extremely easy or difficult tasks so they don't have to be unsure of themselves. They win or they lose. In the end, they miss out on the beauty of the gray area where growth and possibility exist.

Why is this problematic? When people can't handle their partner's periodic requests for privacy or their random bursts of anger and anxiety, they resort to extreme measures (aggression, breaking up) instead of working through normal relationship snafus. When you try to stop your partner from doing things you don't approve of, when you start prying and spying on your partner, when you start avoiding things that make you uncomfortable, then you prevent them from being free and autonomous. Partners learn that it's better to hide and conceal than pay the price of revealing things that might make you uncomfortable.

There is an alternative to this need to predict, be in control, and find closure when it is impossible with an ever-changing person beside us. Reach for the Explore knob. When we have healthy, respectful curiosity, when we support our partner's explorations, relationships are strengthened.

When Normal Feelings of Anxiety Turn into a Problem

Anxiety can become difficult when it is intense and frequent. Sometimes the potential danger to our health and survival is real—like when we come close to stepping on a coiled rattlesnake when hiking a desert trail. Other times, our threat-detection systems can be inaccurate, like an overly sensitive smoke alarm that goes off when you take a steamy shower. There's no risk of the house burning down, but we are probably better off for having a smoke alarm that doesn't take any chances, right? I would argue no. If long, refreshing showers are a harmless, enjoyable part of the day then why should they be abandoned? It's easy to get a smoke alarm that distinguishes between steam and the smoke emanating from fire. You don't get a Purple Heart for suffering needlessly, you just lose moments that diminish your quality of life. If there are other people sleeping in your house when the alarm goes off, then even more moments are lost.

Detecting danger when everything is safe has consequences. For instance, people with panic disorder worry they are going to have anxiety attacks. When they experience minor physical sensations such as shortness of breath or a racing heart, they get anxious about getting anxious, wondering whether a heart attack is imminent. Unfortunately, any number of normal activities are going to raise your heart rate and alter your breathing pattern: jogging on a treadmill, having sex, or eating a burrito topped with spicy salsa.

Worry is about expecting something bad to happen, being concerned about making mistakes, or being judged negatively by other people. Worrying is that incessant mental chattering where we become preoccupied with our own thoughts and feelings. It can help to make a distinction between functional and dysfunctional worry.

Worry can be a very effective problem-solving strategy. If I am walking home alone at 3 A.M., worrying about a sneak attack by a mugger reminds me to stay alert. I check to ensure my wallet is tucked away and move toward well-lit areas.

But sometimes worry doesn't solve anything. I might be in the same situation, and I imagine being mugged, kidnapped, and viciously assaulted, and then I start to scare myself. And if that wasn't enough, I start to worry about what it means to be worrying so much. Is the worry I'm feeling telling me that the danger is real or that I have mental health problems? I spend so much time in my head worrying about horrible scenarios that could have happened that I never had a chance to disconfirm these beliefs, recognizing that the walk was safe. Instead, I might consider myself lucky for getting home. The problem is that the next time I am walking alone I may be even more frightened. I might decide to avoid walking alone completely. This avoidance initiates a dangerous precedent, with the safe, secure space in my life shrinking.

Anxiety is not the problem. The problem is the massive amount of effort devoted to managing anxiety, avoiding situations that made us anxious in the past, and avoiding situations that we think will make us anxious in the future. For instance, you might want to call in sick at work so you don't have to see coworkers who laughed yesterday when you belted out a loud fart in the group cubicle. That's fine if it's one day, but what if you had to give an important presentation to seal a business contract? What if it went beyond one day and by hiding and being ultrasensitive, the fear started eroding relationships with colleagues little by little? Instead of living your life—mistakes and embarrassments and all—you are hidden in the background battling your emotions and your sense of self. Excessive effort and time devoted to altering, controlling, and avoiding anxiety has the unintended consequence of transforming normal anxiety into problems.

Worry Drains Mental Resources and Energy

Worrisome thoughts consume the limited amount of working brain power at our disposal. When we worry, we have less brain power available to take care of other tasks.

Imagine driving on a highway and trying to get to work on time when you're running late. You might feel anxious. You might worry about what your boss will say to you, or you may fret about why you sweat the small stuff and can't get rid of these anxious thoughts and feelings. The unfortunate by-product is that your concentration is not as sharp (maybe you missed road signs for your exit or were unaware how hard you needed to push the brake to avoid an accident). Being anxious drains the energy necessary to be a curious explorer.

Being Physically Tense and Aroused Is Not the Problem

Sports psychologists have carefully examined how anxiety helps or hinders athletes. One, Dietmar Kleine, took a close look at 50 studies on the topic, adding up to a total of 3,589 athletes (from weekend bowlers to members of high school and college teams to elite professionals). He found that being physically tense and aroused has practically no impact on athletic performance. The problem lies with worrying. Athletes who worried the most about being judged, doing poorly in competition, and trying to control their anxious feelings fail to perform to their ability. I think the reason that you see a problem in performance is that athletic competitions require people to act quickly, and attention is critical (baseball players standing on first base have a matter of seconds to decide whether to steal second base before the pitcher throws next).

The difference between higher- and lower-performing athletes appears to be the amount of brain activity going on when they are doing nothing much at all. Lower-performing athletes show extremely active brains. This might sound good until you consider that the person with brain capacity to spare is in a better position to recognize what they are up against on the field, court, course, or track so that they can perform to the best of their ability. We need room for our brains to focus on what is happening in the present because no two athletic competitions are alike. Depending on the amount of rain on the soccer field, what we ate for breakfast, and how much we stretched before running, our body is going to work slightly differently than at any other time—we are going to have to account for these unique circumstances to nail our best possible performance. Athletes who worry are not only performing worse, they are putting in more effort and don't have the energy left over to be interested and enjoy themselves (the fun part of being "in the zone" is gone). Whatever enticed them to play is being lost because no matter how you look at it, they are failing to reap the benefits.

Nothing is unique to athletes, the same goes for workers, supervisors and managers, parents, friends and lovers, and so on. When people worry excessively, they are less flexible in their mindsets and show greater difficulty with back-and-forth shifts in thoughts, actions, and tasks (multitasking). For instance, a worried hiker is going to report less awareness of the scenic beauty and fewer minutes spent totally immersed (the joyous sense of oneness with the mountain).

Physically and Emotionally Drained for Whatever Comes Next

Intense efforts to regulate anxiety exhaust mental and physical resources. Afterward, a person operates far below full power for any new challenges that arise.

Let's suppose the boss tells you that you need to meet about something important at the end of the day. Now, you may ruminate about the meeting and start preparing for the worst as you periodically check the time. You may also worry because your normal routine is to head home straight from work to watch the kids so your wife can get a break. All of that effort to control your anxiety is going to leave you with little energy to effectively play with those kids. If you struggle with worry, then you're going to be physically weaker when your kids ask you to carry them. You are going to be mentally drained and actually show less intelligent thought if you want to help them with homework.

Overtaxing a person's resources is destructive. When we feel extremely anxious and aroused, our attention narrows. Trapped inside our heads trying to avoid and control anxiety pulls us away from the present moment and further away from being successful at our goals.

Dozens of experimental studies show that if you ask people to engage in intense acts of self-control, such as trying not to appear anxious when giving a speech in front of an audience, they are less effective at responsibilities immediately after—even if the follow-up activity has nothing to do with the first task. In studies, they score lower on intelligence tests, show a weaker hand grip, hold their hands in cold water for less time before giving up. They are less able to resist eating unhealthy snacks or staring at pornographic pictures, and they are more susceptible to compulsive purchases when shopping. Taxed by worrying, we essentially become psychologically and physically weaker. This research is relevant to all of us. Each of us has moments when this happens; we just differ in how often or intensely it occurs.

With Too Much Anxiety, Chaos and Stagnation Reign

The research is clear that very anxious people and people who tend to avoid anxiety or get overly worked up about feeling anxious are worse off than their peers. By fearing anxiety, a person can't act on their curiosity. If overwhelming worry affects us early in life, then anxiety can disrupt a lot of opportunities during a critical developmental period.

My friend's 16-year-old daughter Sonja suffers from dyslexia. Sonja's condition requires her to expend enormous amounts of energy to make sense of information she gathers from the environment. Even everyday tasks that are effortless for her peers can leave her fatigued. Sonja is so exhausted from navigating everyday life that learning at school, listening carefully during conversations, or watching a movie often come to a grinding halt. By the end of the day, she feels overwhelmed by sounds, colors, and people and shuts down to limit stimulation and release the bubbling tension from the day. Sonja's dyslexia lowers her threshold for being overwhelmed. She starts feeling swamped a lot earlier than family and friends do.

Unfortunately, this means that dealing with tension and anxiety often takes priority over being interested or satisfied. To cope with these difficulties, she carefully manages her environment to chisel away bits of anxiety far in advance. She rehearses conversations because her brain doesn't always cooperate during spontaneous, pressurized moments. She doesn't like restaurants if she doesn't remember the name and location. Friendly's is a safe haven; she knows the menu and the familiar layout inside. Structure keeps the painful feelings of anxiety and doubt at bay. But for a long time the price was high—her curiosity was forsaken. Her rigid need to control her environment limited her zest for life and her ability to overcome her dyslexia. Fortunately, with the support of her parents and specialized schools, her laser focus on maintaining her equilibrium is easing, and her world of learning and friends is beginning to expand.

Sonja suffers from dyslexia but her story is also characteristic of people who suffer from severe anxiety. At the extreme, approximately 16% of people living in the United States have been diagnosed with an anxiety disorder at some point in their lifetimes. This makes anxiety the most common psychological disorder in the population (more common than depression or substance use). Sufferers of anxiety disorders devote their lives to managing anxiety at the expense of working toward more important, enjoyable, and meaningful goals. Making life choices with the function of avoiding and concealing anxiety establishes an insulated and depleted existence. I call this living in a social cocoon.

Among the many different anxiety disorders, social anxiety can be most devastating to our well-being. Think about it: The most meaningful and desirable part of life, spending time with other people, is the source of greatest dread. Being in a social cocoon keeps a person safe from anxiety at the expense of residual, unsatisfied curiosity, passions, and sources of meaning. These unsatisfied desires tend to linger, sometimes for decades, with the socially anxious person ruminating about what might have been.

I worked with many clients suffering from social anxiety disorder. I remember a running back for one of the best college football teams in the country. He was a good-looking, well-mannered, and kind guy. But he also suffered from intense social anxiety, not something you expect in a hulk of a football player. Playing sports was possible only because he was so damn good that he had a huge entourage that protected him from anyone who so much as pointed at him (so he never had to be assertive on his own), took tests for him (so he never had to learn how to handle the pressure of being called on), brought girls he liked to him (so he never had to develop dating skills), and made sure he was never left wanting for anything.

When he graduated college and his entourage moved on, his social worries took over his life. All he wanted to do was mentor and coach inner city children in sports. But he couldn't handle his worries about what other people thought of him. Playful banter in the teacher's lounge was foreign to him. Easily intimidated by the principal and students, his fear of being judged led him to quit. Next thing you know, he's in the unsatisfying position of hauling furniture for a moving company. Finally, he had the courage to respond to an advertisement for a study we were doing on social anxiety. This was fortunate because most socially anxious people never seek treatment. Why? Sadly, the same fears of being rejected by other people cause people to fear being stigmatized for seeking treatment.

In an initial session, the typical client suffering from social anxiety describes an intense fear of being rejected by other people. When they were younger, they didn't act like normal teenagers. They didn't go to parties, they didn't date, and they didn't play sports or join clubs. They stayed home on weekends not because they wanted to but because they were afraid of being an idiot, not being funny, not knowing what to say, and just feeling unlovable. Social fears led them to be isolated and lonely. The worst part is that these behaviors led them to be less prepared for the adult world. They missed out on those practice years where people learn social skills like how to make good conversations, be witty, and figure out how to kiss.

Anxieties aren't always visible on the surface. Many of my clients appeared to be doing fine. Some of them are successful engineers or accountants with six-figure salaries, families, homes, and two cars in the garage. But pull back the curtains and you realize that their anxiety has derailed them. They may have lowered their goals, narrowed their aspirations, and avoided situations that made them uncomfortable as life became less meaningful and satisfying as the years went by. Dreams deferred.

Anxiety Is Universal but Our Attitudes Toward It Are Not

Although we might infer that highly curious people are simply less anxious, my research tells a different story. The most curious people weren't merely going through life with little concern about being judged and criticized by other people. They feel just as anxious as less curious people do. What was different was that they acknowledged their feelings as being normal. Whether anxious or joyful, curious people don't hide from their feelings. They show a tolerance for distress.

A healthy approach to living is all about psychological flexibility. Intense feelings of curiosity are acted upon and the world is explored, and for reasonably dangerous activities, anxiety is given careful consideration as to whether escape is the most viable option. Unfortunately, for people with anxiety problems, escape routes end up being the default reaction to anything recognized as novel, challenging, or uncertain. When it comes to social anxiety problems, too often that Anxiety dial is cranked to 10, while the Explore dial is ignored.

Over time, an accumulation of unfulfilled desires and rewards has considerable psychological costs, from regret to hopelessness. We know from studying people's retrospective accounts of how they lived that suffering from actions not taken is more intense and enduring than taking risks and failing; an important lesson for when each of us is at the crossroads of risking or retreating.

If we don't act on our curiosity, we don't get any of its benefits. Worrying often prevents us from acting on our lust for the new. From my work as a therapist and scientist, I suspect that the lost discoveries, accomplishments, and dreams of people failing to act on their curiosity are staggering. It's hard to calculate the costs of unsatisfied curiosity. Who knows how many people lose out from never being the recipient of a person's untapped strengths and potential.

When you worry about failure, you don't take risks. Without those risks, cures for diseases would have never been found, artists would have never shared their work with the world, and great leaders would have never had the courage to voice their ideas.

Searching for the Elixir of Optimal Anxiety— Butterflies Are Best

If we are going to understand how to make anxiety work for us, it starts with self-awareness. Too little anxiety or challenge sets the stage for boredom and apathy. A moderate dose of anxiety is just right, harnessing our energy and attention to maximize our potential in an efficient manner.

I asked Andy, an emergency room surgeon, about how anxiety plays a role in his work. When Andy doesn't feel anxious at all, he has no problem doing a routine surgery, say, removing a few kidney stones. He is on virtual autopilot, able to banter with nurses, sing along to Radiohead songs playing in the background, and still do fine. Yet, there is a price for this mindset. Without anxiety, he isn't geared up to be at the top of his game. Relying on his expertise instead of a feeling of urgency in the moment, his attention can wander. If there is something amiss, he is more likely to overlook it. This makes sense because we know that when something unexpected occurs, the unchallenged person is more likely to miss it. Almost assuredly, the patient gets a lackluster performance.

When Andy is too anxious, he is so self-focused on doing the right thing that he has considerably less attention remaining to get out of his head and into the operation. The ideal amount of anxiety is somewhere in the middle—not too low and not too high.

The ideal is when we are mildly anxious, with butterflies, going into a situation. This bit of anxiety energizes us, directing our attention to the details of what needs to be done and with a deep breath, we buckle down and are attuned to the unique moment we reside in. Our anxiety keeps us vigilant of multiple goals. If you are a surgeon like Andy, perform the least invasive procedure for the problem, monitor the patient's vital signs, and be on the lookout for anything unusual that requires a change in plans. Optimal anxiety ensures a perfect combination of carefully avoiding errors and eagerly searching for opportunities to improve things further.

The right amount of anxiety unleashes our strengths and skills so that we perform at our best. When we aren't working on anything in particular, a moderate dose of anxiety betters our moments. As mentioned in Chapter 2, even though people think they want to know why a stranger was nice to them, greater, longer-lasting interest and meaning is experienced when it's random and mysterious, when we have a hint of butterflies, of the unknown.

Getting to that optimal level of anxiety where we are a curious explorer, engaged and enjoying ourselves, depends on what we think about situations we find ourselves in. Nearly any task can become interesting if we perceive an activity as being novel, challenging, or mysterious.

One of my students is a waitress at a restaurant in Yorktown, Virginia. Faithful, high-tipping customers head over straight from work. There's nothing unusual about this except that the closest job site isn't a shopping mall or a coal mine, it's a battlefield. A slew of adults in Revolutionary-era regalia eat at the restaurant during breaks from reenacting the Siege of Yorktown (this was the battle that ended the American Revolutionary War in 1781; trust me, I didn't know this until she told me). Walking past a full regiment of Continental soldiers dressed in battle gear and lugging mock cannons on a 100-degree summer day might be novel and interesting to you and me but much less so to her.

Novelty alone doesn't make you curious. Suppose that without knocking, you walk into your boss's office, and he is sitting at his desk, completely naked, typing away on his keyboard. You ask yourself, *what the %$#^@?* Clearly, you are being bludgeoned with novelty (I'm assuming this is not a normal occurrence at your job). You might giggle, wondering what is going on—at work, no less. Or, you might feel violated, repulsed. People are going to respond to situations differently, and some people will not be curious even though, objectively, they find themselves in a novel, mysterious, surprising situation.

If we want to understand what makes us curious and anxiously alert with butterflies, clearly something else is involved besides the novelty or desirability of a situation. Two factors are at work—novelty potential and coping potential.

Novelty Potential—Judging an event as new, challenging, unexpected, complex, surprising, mysterious, or challenging means that we recognize opportunities to learn and grow.

Coping Potential—Judging whether you have the skills and abilities to deal with or make sense of the novelty being confronted.

If you put these two beliefs together, you can see how some workers might feel anxiety and disgust upon walking in on their boss in the buff (high novelty, low coping) and how some workers might feel curious and amused (high novelty, high coping). Remember, you are judging an event—there is no good or bad response; it's what you think and how you feel.

For all of us, there is a point when tasks become so confusing or difficult that they become frustrating and are no longer interesting or enjoyable. Consider a leisurely day at an art gallery with friends. If you're like me, a novice when it comes to art appreciation, you spend less time viewing paintings and sculptures than do experts who scrutinize details that I fail to even see. When a piece of art is difficult to understand, we novices are less interested. Because it's so easy for us to move on from one gallery to the next or to space out, our frustration is unlikely to last too long. Escape is easy. Few people go to art exhibits, listen to music, or read books during their free time to feel overly frustrated and anxious. They're looking for dividends for their efforts. If the rewards are too small, the discomfort isn't worth it, and people abort their plans and go look for something more interesting and enjoyable.

Interestingly, when challenging and complex artworks are titled, the artwork becomes more appealing to the average Joe and Jane going to the museum. Titles offer an entry point into the previously impenetrable mind of the artist. By reducing the confusion a tad, the artwork becomes more meaningful, dropping us down a notch from feeling too anxious to that zone of optimal anxiety. Now, if you happen to be extremely knowledge-able about art, regularly reading art magazines and jet-setting around the world to go to museums, you have less need to rely on titles. You relish being stumped. When you are an expert, your coping potential is so high that you need a vastly tougher challenge to feel the butterflies (too often you are pained by too little anxiety).

Whether it's art, music, sports, conversational topics, or foods—whatever it is—if you view something as new and challenging, you are going to be more engaged and feel more joy and pleasure than if you were in a situation that didn't make you anxious. When we enter new territory with no confidence that we can understand or do well at what we are do-ing, frustration and anxiety dominate and whatever interest and excite-ment we initially had are going to be obliterated. If we are challenged in ways that force us to stretch our skills and knowledge to their maximum, we are going to experience intense curiosity. If we avoid challenges where some anxiety is present, we will be fragile, overwhelmed far too often, and devoid of curiosity too frequently.

The Right Fit—
High Novelty, High Coping, Feeling High

We become friends with our anxious feelings when our coping potential matches the novelty of a particular experience. Moreover, our competence builds one situation at a time. We gain confidence and competence for the next uncertain, challenging situation that we face.

You can see this change in Amy as she dealt with a condition called synesthesia, in which there is a mismatch of typically separate reactions to being stimulated, as when someone hears colors, sees flavors, and smells letters and numbers.

When I was five and found out my mom didn't know what red sounded like, I was floored. By the time I was 17, I thought I was crazy. I used to mention my syn—wow, that man is blue—and my friends used to tell me to be quiet. So I knew other people didn't see colors around people. I used to get embarrassed about it. I tend not to mention it to people unless I trust it isn't going to change their opinion of me. I had experiences where I told someone and got laughed at. Being told you are a liar wears you down. It can be isolating.

Amy struggled with her condition because it could be a stimulation overload. For Amy, taking a shower as a child was terrifying. Now, as an adult, showers are still anxiety provoking when she is in the wrong mindset. Barely awake in the morning, water hitting her body, steam in her face, lathering soap in her hands—all these experiences flood her with colors, sounds, and images that are hard to handle. Like an intravenous dose of coffee, it shocks her nerves.

Then she opened up to a boyfriend and asked him what he sees and hears when music plays. She realized that her brain offers her a much more interesting palette than the rest of us. Incoming information produces pictures in her head. She sees sounds as colorful textures, and colors make noise and have flavors. "Nice" color palettes sound like children giggling. "Ugly" color palettes sound like a bad orchestra warming up or screeching violins. Listening to Tool (an excellent metal band, at least to this critic's ears), she sees jagged orange and turquoise waves that rapidly swirl around, akin to a spiral staircase. Realizing her uniqueness brought a new attitude toward life. She began cranking her Explore knob as far as it would go.

When she relinquished the battle and observed her experiences, she was surrounded by undetected, underappreciated meaning. I suspect many of us do the same. We can reclaim our moments with a new mindset of mindfulness and self-compassion, recognizing novelty and building our ability to embrace it.

If we want to reach this optimal state where anxiety, growth potential, and competence meet, then we need to know what is interesting and challenging to us, and we need to recognize that some situations are going to provide the optimal feeling of butterflies.

With a colleague, I studied nearly 500 adolescents in Hong Kong to figure out what happens when highly curious kids experience "good fit" and "bad fit" with their high school. We figured that very curious kids thrive in school—with good classroom grades and excellent scores on national achievement tests—when they believe their school challenges students to gain mastery and knowledge (high novelty and high coping potential). This is exactly what we found.

While very curious kids might be anxious when they are challenged with high expectations, they are also more engaged, do better in school, and report better relationships with their teachers. The very curious kids thrived compared with their less curious schoolmates. They need high novelty situations to hit that optimal zone of anxiety or they are underwhelmed. On the flipside, very curious kids were worse off than less curious kids when they believed their schools failed to intellectually challenge them. They didn't go to class because they didn't need to study hard to ace exams. Their strengths were underutilized and in response, they mentally dropped out.

We need to attend to what interests us as well as what makes us worried. Even though very curious students initiate three times as many classroom questions compared to their less curious peers, both groups become silent when teachers are viewed as threatening. This saps their energy so there isn't enough left over for homework, after-school activities, and socializing.

I suspect the same goes for you and other adults in the workplace. If you don't feel safe, if you don't feel there is respect and interest in your values, you tune out and drop out. When you do, others are affected by your declining performance. There is much to learn about what an environment can do to elevate or squelch our greatest assets.

As another instance of the power of fit, think of parent-child trips to the playground. Watching a child's conquests and interactions with other kids can be a blast of surprises. Novelty potential is high, and the attentive parent is often curious. As any parent knows, however, playgrounds also leave a knot in your stomach. Fun remembrances of diving head-first down the slide and soaring across monkey bars dramatically change when the roles reverse. Now you're trying to protect your kids from eating dirt, chewing on another child's skull, or chasing a lunatic ice cream truck.

Traditional playgrounds usually have blind spots so adults have to be on the move, watching and calling kids who dart out of sight. When parents are on guard, worried about the safety of their kid, trying to manage anxiety, they are liable to lose that joyous curiosity. When parents get too anxious, visualizing injuries before they happen, they don't let their kids be independent and have fun, and everyone loses out. To get the most from the experience, you need to feel that you can handle and relish the novelty.

In McLean, Virginia, there is an 18-acre park called Clemyjontri Park. This remarkable park was designed so children using wheelchairs, walkers, braces, or with developmental disabilities can play alongside those without disabilities. Design features include wider openings in the play structures, monkey bars that are lower than usual, ramps that connect structures, and a rubber ground surface so kids can fall off the rock-climbing wall and bounce back up without gut-wrenching screams.

Robin and her two-year-old daughter Lilah went to Clemyjontri Park with a play group, welcoming the unintended benefits of the design.

Parents don't have to run around structures or look up tall ladders to keep an eye on their kids. Instead, they can watch their children play from where they sit. As a result, Lilah could play with other children for a longer period of time without Robin's interference, and Robin could observe aspects of Lilah's personality that she hadn't seen before. For instance, Lilah took on the role as leader of her posse of 18-month- to 2-year-olds, pointing to new places for the other children to "join her." Robin could also enjoy stimulating conversation with other moms. In the right situation with the right level of anxiety, everyone benefits.

If you want to capitalize on strengths and maximize well-being, pay close attention to the novelty being offered and whether there is an opportunity to handle and cope with ongoing challenges. The right fit can activate curiosity. Threatening situations dissolve curiosity with a rapid shift from wanting to explore to protecting the self.

These findings on fit have important implications for creating optimal environments at work, play, or any organized activity. Strengths need to be clarified. Name them. Acknowledge them in yourself and others. Beyond the basics such as courage, generosity, and curiosity, there are strengths to be found littered in everyday acts. The person who can make the most complex ideas sound simple (the decoder), use few words when others require sentences (the sharpener), energize others in proximity (the catalyst), and allow others to showcase creativity and strength that they didn't know they had (the incubator). By giving them names, strengths are nourished. When we are granted the autonomy to use our strengths, we are more engaged, more invested, and we perform better. When we feel constrained by rules and regulations, it's easy to flounder. Consider curiosity in the school system. If a school cares more about raising student test scores to gain prestige and funding as opposed to whether kids actually gain knowledge, then that school is liable to squelch students' natural curiosity, leaving them disconnected and underperforming.

We can find out if people need more structure and guidance—or less. We can check in to see if more advice, more quiet, or more challenge is needed. A very curious student/worker/parent might need a bigger, more dramatic goal; the less engaged might need more excitement to trigger their interest, and the very anxious might need additional reassurance, information, and comfort so their curiosity isn't depleted by worry.

Strategies to Live Fully with Anxiety

The premise that anxiety is bad and harmful and that it must be eliminated is an idea that has outlived its time. Therapies to eliminate anxiety shouldn't be trusted. Scientifically informed techniques can help you form better relations with your anxiety and not get caught in a death-grip struggle with worry and wear yourself out being anxious about being anxious.

Strategy 1: A Survival Guide for Quicksand

We've talked about dialing up the Explore knob to begin aiming more and more of your actions toward the direction of your stated values instead of futile efforts to control anxiety by doing more of the same. With the goal of starting to think about working with instead of against anxiety, I want to quote another metaphor from the work of Steven Hayes. He calls it the Quicksand metaphor. Why the metaphors? By meditating on them, we can get out of our heads and shift perspectives to focus on new strategies and face old problems.

> Suppose you came across someone standing in the middle of a pool of quicksand. Without ropes or tree branches to reach them, the only way you can help is by communicating with them. They shout, "Help, get me out," and begin to do what people usually do when they are stuck in something they fear: struggle to get out. When people step into something they want to get out of, be it a briar patch or a mud puddle, 99.9% of the time, the effective action to take is to walk, run, step, hop, or jump out of trouble.
>
> This is not so with quicksand. To step out of something it's necessary

to lift one foot and move the other foot forward. When dealing with quicksand, that's a very bad idea. Once one foot is lifted, all of the trapped person's weight rests on only half of the surface area it formerly occupied. This means the downward pressure instantly doubles. In addition, the suction of the quicksand around the foot being lifted provides more downward pressure on the other foot. Only one result can take place: the person will sink deeper into the quicksand.

As you watch the person stuck in the quicksand, you see this process begin to unfold. Is there anything you can shout out that will help? If you understood how quicksand works, you would yell at the person to stop struggling and to try to lie flat, spread-eagled, to maximize contact with the surface of the pool. In that position, the person probably won't sink and might be able to logroll to safety.

To the person trying to get out of the quicksand, this is an extremely counterintuitive idea. Someone struggling to get *out* of the mud may never realize that the wiser and safer action to take would be to get *with* the mud.

If you're like most human beings, too often you get stuck in the quicksand of your own anxiety and try to thrash your way to a resolution while bypassing the optimal anxiety that adds excitement and leads to peak performance.

If you want to know what it feels like to be stuck in the quicksand, go to a quiet place where no one will be around so you can get into the following exercise.

Okay, I am going to ask you not to think about something. Ready. Here it comes. Now remember, don't think about it. Don't think of . . . warm chocolate cake. You know how it smells when it comes out of the oven. Don't think about it! On the outside of the cake, there is a moist, warm layer of chocolate frosting. But don't think about it! Think about anything you want, just don't think about the cake. Close your eyes for a minute and whatever you do, don't you dare think about that warm, moist, frosted cake.

Did you think of it? How hard was it to move your mind to something else?

If you are like most people, you thought about the cake. This is because I put the idea into your head and then basically asked you to ignore it. That's the irony of trying to hide, or suppress thoughts, feelings, or images. Trying not to think about something binds us even closer to it. We create a rule in our mind that says, "I refuse to think about that email where my brother called me a prick." Now, every time you try not to think about it you are forced to go over this rule again, which leads you to think even more about it.

I spent a year working at the Veterans Administration Hospital in Charleston, South Carolina, treating combat veterans traumatized by their war experiences: guns being fired at them, carrying injured friends in their arms, and witnessing the limp bodies of civilians and innocent children. They couldn't shake these images, and the harder they tried, the more vividly they returned, interfering with ordinary activities like falling asleep or listening to their child tell them about school. If you have ever tried to avoid thoughts because they made you feel uncomfortable or tried to escape (through drugs, work, exercise, sex, collecting porcelain figurines of hippos, you name it), you have something in common with these veterans. What do you do? To begin the change process, I want you to meditate on this metaphor of relating to your unwanted private experiences in a new way.

Imagine that you are in a swimming pool with an inflated ball in your hands. The inflated ball is the thoughts or feelings you don't want. So, what do you do? You try really hard to push it under the water so you don't have to look at it anymore. Now, it's hard to hold an inflated ball underwater. Isn't it? If you don't put all of your energy and focus into it, that ball is going to burst right back to the surface. Now, as long as you put a lot of energy into it and you aren't distracted, you can keep that ball underwater for a long time. Unfortunately, you rarely ever get the pool all to yourself. From the corner of your eye, you notice several beautiful women in tiny bikinis with perfect bodies smiling at you as they walk toward the pool (or gorgeous men with washboard abs and sculpted biceps). What happens? You forget what you were doing, that ball shoots up to the surface, and there it is—that thought you've been trying so hard to get rid of. And so you submerge it again. You look down because you can't watch and flirt with those beautiful people *and* keep the ball under water. And then the most beautiful of the lot jumps into the pool, swims over, and starts talking to you. Now you're stuck, and you're getting worked up about it. You don't want to have that thought, you fear having that thought, especially now, and yet, you really, really want to talk to her. Not only are you tense and tired from holding the ball underwater, now you are tense from trying to socialize at the same time.

What's the solution? Take a close look at that inflated ball. Let it rise to the surface of the pool and touch it, stare at it, smell it, taste it, and keep doing this for as long as possible. And then when you go to the pool tomorrow, with the ball, do the same thing. It doesn't matter if you bring the ball or if it just appears; if you see it, focus on it. And I don't mean let your eyes glaze over like a zombie, sleeping with your eyes open. Focus on the details. Look at the colors, feel the surface where it's smooth and where it isn't, turn it in your hands, and see how it looks really close to your eyes and how it differs when you hold it at a distance. Do this day after day and what do you think is going to happen? You aren't going to get anxious. You are going to be bored senseless. What do you expect? You're going to stare at the ball for an hour at a time, several times a day, for months. And this is the goal: to get bored, really bored. And what will happen? The ball, the thought, will still be owned by you, one of many possessions in the cluttered garage in your mind. It just won't have the emotional sting it once had. You aren't going to feel a need to hide it or get it out of sight. You are going to accept it for what it is, for all the qualities it has, and because of this, you are going to be free. You aren't going to feel compelled to keep it under the water any longer because looking at it is going to be just like any other thought or experience. You are going to be able to swim, enjoy the feeling of water against your body, talk to people who catch your attention, and gravitate toward what truly interests you.

The goal of this mindset isn't to be less anxious. The goal is to be alive and reclaim moments so you build larger and larger behavior patterns that are aligned with your values. Take the tension with you and choose to move in the direction of what matters, where meaning lies.

Strategy 2: Defusing Nasty, Frightening Thoughts

Going back to what makes us curious, we might recognize novelty, challenge, and uncertainty, but because we don't think we have the skills to cope or comprehend, we get too anxious and leave situations to do something that makes us feel calmer. Perhaps we end up in a job far beneath our education and training, we end up in a relationship with someone who doesn't interest or challenge us, or we end up bored in our free time because we are unwilling to try new things that might make us look silly. Don't think this pertains to you? Because of these thoughts, people often avoid getting out on the dance floor, and if they do, they don't let it all hang out, they play it safe, snapping their fingers, softly clapping their hands, swinging their arms just a bit, moving their hips just a bit, nothing that might bring attention. The result? Challenges are kept to a minimum, lessening our chances for finding meaning. We can handle and comprehend more novelty and challenge than we think. What keeps us bored or anxious most of the time are thoughts that we accept literally.

Critical to working with anxiety is recognizing the power of words. What's problematic is when we treat our thoughts as if they cannot be questioned or tested and are, instead, the truth. Your friends might ask you to go mountain biking with them on an off trail through dense woods, something you have never done before. Your mind starts racing: "my friends will laugh," "I can't do this," blah blah blah. If you get caught up and buy into these thoughts you are doing what psychologists describe as "fusion." Fusion is about viewing thoughts as the literal truth ("I can't do this") instead of what is actually happening ("I am having the thought that I can't do this"). Fusion makes it nearly impossible to be in the present moment and act in ways that fit with cherished interests and values.

We get tangled up with our thoughts more often than you think. We think "I'm not funny" because nobody at dinner laughs at our comments. What do we do? We talk less, becoming reserved and lonely at a table full of people. Our thought causes us to shut down. If it's not this thought, it's another one that you fuse with. Take your pick: I have nothing to say, my boss doesn't like me, I am not interesting, people are going to think less of me if they see how anxious I am, I should be working harder, and so on.

Most of the time, we don't realize that we structure the world according to what our mind tells us. We assume our thoughts are accurate reflections of us and what is going on in the world. When our anxious thoughts push us away from instead of toward the people, things, and places that trigger our interest, we have a problem. Fusing with thoughts gives them extra power and weight that they don't deserve.

There are several strategies for changing how we relate to our thoughts. Give me one minute of your time to give you an idea of how you can defuse uncomfortable thoughts.

I want you to think about milk. Just those four letters together: M-I-L-K. Nothing exists except for the word MILK. What comes to mind when you say the word out loud—MILK?

Don't force it. Just let it all come to you. Perhaps you picture a creamy, white fluid being slowly poured into a bowl of cereal. After drinking a glass of it, you might imagine licking a cold, watery streak left on your lips. Do you have a clear image? Good. Now, I am going to ask you to do something a little strange. It's brief, so bear with me and follow these instructions exactly.

I want you to look at a clock, and for the next minute, I want you to repeat the words, "milk, milk, milk" over and over again. Don't say it any faster than you normally do. When the next minute begins, start saying milk. Ready? Go.

What happened? Did the word start to lose meaning? I bet it started to sound weird, the sounds blending together so it felt like you were talking gibberish. I bet you stopped seeing milk cartons. I bet you stopped seeing white, creamy, liquid being poured into a glass. I bet your brain stopped creating images and stopped creating thoughts about those four little letters.

This is what defusion is about. The meaning and impact of the word falls away. Words can lose their power, and you can see thoughts for what they are, just an arrangement of sounds and words. Now, the word *milk* isn't threatening to most of us. But we can use the same exercise with thoughts that do make us anxious. The exercise works best when you can distill your thoughts into a few brief words or a phrase. For example, it you are worried that people won't think you are intelligent, you can use the word *stupid*. If feeling anxious makes you uncomfortable, you can use the phrase, "I'm getting anxious." Just as we did with *milk*, say the words or phrase over and over for a minute, and notice what happens. Let me share a personal story about using defusion to create space for being more interested and engaged.

If there was one thing I was extremely ashamed about growing up, it was my body. My body collected birthmarks and moles, constantly surprising me with a new hairy mole on my arm or an ugly, bulbous multicolored splotch on my chest. In my dreams, there was a mother mole, and when I found and slayed it, the proliferation of moles would finally end. Unfortunately, I was never able to slay the beast, and multiplying moles became the story of my life. It's not flattering to be called "the chocolate chip cookie" when your shirt comes off. So I stopped taking off my shirt at the pool, at the beach, even when it was dragged down by 3 pounds of sweat as I walked the streets on a summer afternoon. My shirt always stayed on because I was worried that people would laugh and stare. And then one day I was at the beach with a bunch of laid-back surfers. One of them turned to me and asked me why I always keep my shirt on. and when I told him the story of my woes, he calmly said, "Now how weird would it look if you didn't have any marks on your body? No scars, no nipples, no moles, no sign that anyone lived in that skin of yours? It brings color to a person. It makes you unique." There was freedom in this statement and I took off my shirt. It hadn't occurred to me until then that for far too long, I had become my moles. I became the ugliness that I saw in those moles. I never even really looked at those moles, I just saw self-loathing. Well, that surfer's words severed the tie. It was rather abrupt, more abrupt than other attempts in my life to work with anxiety and insecurities. What happened in that moment was that I saw my thoughts for what they were. It wasn't that I was ugly, I was "having the thought that my body was ugly." By weakening my attachment to those thoughts, by weakening my belief that it was true that my body was odd and hideous, the thoughts became less important and had less of a hold over my well-being and what I wanted to do. I even went through a period where I was determined to be shirtless as often as possible in as many places as possible. As a teenager, I never took my shirt off; in college, I almost never kept it on. Gates were opened.

This exercise isn't the only thing you do to work with anxiety. What I want is for you to experience new ways of playing with your anxious thoughts and feelings. View your thoughts from different angles. Picture them as conversations with your mind. You can be interested and open to what your conversation partner, your mind, has to say, but you don't have to agree with him or her. Thank your mind for the unpleasant thought and continue moving toward things that are interesting and meaningful. Picture your thoughts as animate objects. "I shouldn't have made so many mistakes" becomes a purple, puffy, La-Z-Boy chair. "That success was a fluke" becomes a pink turtle. Choose your favorite image. It becomes a lot easier to confront unwanted experiences and to prevent fusion when they look silly and nonthreatening. Defanged snakes. Toothless pit bulls. Stingerless bees. They look nasty, but the threat, the barrier, is an illusion, nothing more than a product of our brain. We can observe this brain residue with an overstuffed burrito of mixed emotions (anxiety and curiosity) as we move onward.

Strategy 3: Choosing a Direction for Your Life with a Well-calibrated Compass

You are not your thoughts. Your thoughts are not an accurate reflection of what you can do. Your thoughts are not an accurate reflection of the risks and rewards out there. But if you are not going to trust thoughts and feelings, then what is going to help us decide what to do? The answer is curiosity, interests, and values (see Chapter 5). Start searching with a few brief exercises. All you need is a pen, paper, and some time for reflection.

Picture yourself 20 years from now, having grown wiser and more compassionate with decades of added knowledge and experience. Spend some time fleshing out what this future self looks like. What would you be doing and what type of person would you be if you kept on doing the same things you do right now in your life? What happens if you keep on this same track for the next 20 years? What are your good friends saying about you? What are you doing for a living? What have you accomplished? What have you failed to accomplish that you wanted to? Get a good image in your head.

Imagine what this "future wise you" would say to the "you" that is alive, here and now, in the present. They might offer you some perspective on what's working and what isn't. Perhaps their suggestions for getting the most out of life are very different from how you act now. Perhaps you are on the right path. I want you literally to write a note to yourself. What advice does this "future wise you" have to offer? Write it down.

There are other ways to consider whether you are deviating from what interests you. I want you to sit with the idea that there are costs to ignoring the curious explorer within you. Thinking about your life trajectory might motivate you to do something different now instead of waiting for some earth-shattering event that might never happen. Here's another exercise to try.

You are going to be dead a long time, and only have a brief period of time to feel, to do something meaningful, to do something that matters. Imagine your own death. Even though we might not like to think about it, there is nothing we can do to prevent it. It's inevitable. Picture the engraving on your tombstone after decades of sticking with your current routine. On your tombstone is a single paragraph about the life you led. It talks about your personality, your accomplishments and contributions, your missteps and failings. What's it going to say? If you keep doing what you're doing, are you going to like what it says? What don't you like? What do you like? What becomes your life story? We are talking about your legacy, your mark on the world.

Of course, these exercises aren't the only way to mitigate anxiety by bringing interest and openness to the forefront. As a final testament to this new perspective of living now instead of waiting for the pain to be managed, of being driven by curiosity instead of fear, let me share the story of one of my most memorable clients.

Rick, a teenager with intense fears of insects, dogs, cats, dolphins, and more, dreaded the spring and summer months. Other kids in school would visit each other, play outside, and take part in sports and weekend trips together. Rick had a hard time making friends because he rarely left home. When he went outside, he didn't see roses and chrysanthemums, he didn't see children playing hide-and-seek, he didn't see the beautiful girls in sundresses, all he saw was an army of bugs that could ambush him at any moment. Unrelenting fears took over, and in his brain was a blackboard with the same words written over and over again, "stay alert for bugs, they're quicker than you, be ready to run." Anticipating the worst—bugs crawling on him, leaving sticky fluid on him, and maybe even biting him—he spent several months each year in hiding.

A 13-year-old with a fear of butterflies is an invitation to being bullied relentlessly. Sadly, he was. So, we began working with his fears. I realized his strength was his love for learning. He came to sessions talking about Darwin's theory of evolution, imagining how time travel might be possible if humans can journey through black holes, and so on. I realized that I could use his passion for science to view animals and insects differently. That is, expose him to his fears but instead of trying to eliminate his fears, change his goals. In order for us to get to what we value, we have to take our anxieties with us.

We started slowly, reading books about butterflies, watching movies about butterflies, staying in a room with a butterfly in a jar, having him touch the jar, and eventually holding and looking carefully at the butterfly. I was ordering butterflies from tropical dealers so that we could look at how they evolved different features to survive their local conditions. We made a game out of it. We would read about the local climate, and when we released a butterfly, he had to figure out what features would show support for the idea it evolved and what features didn't. He looked forward to therapy. He was anxious and in the beginning, when I brought a butterfly in a jar into my office, I found him flattening his body against the back wall. But as we brought this game into the picture, he spent longer and longer periods of time with butterflies and nothing bad happened, his Anxiety knob went down and his Explore knob was turning to higher levels.

Then we took it a step further, and started visiting a butterfly conservatory two or three times a week. We talked about what he was worried about and what he was excited about. On the first visit, he freaked out. Hundreds of butterflies, dozens of varieties, flapped against our exposed skin and landed on our clothes. He screamed, and since he hadn't hit puberty yet, it was a high-pitched, blood-curdling sound. And he ran as he screamed, arms flailing in the air. Looking terrified and fragile, I began to see why other teenage boys might bully him. So, we returned to the idea of working with his passions and interests—and taking his anxiety along for the ride.

He told me that he wanted to be a scientist when he grows up but wants nothing to do with animals or insects. I asked how he planned to avoid these living creatures? It became clear that he couldn't. The next time we went to the conservatory, I planned a visit to the nursery where the entire life cycle of the butterfly could be observed, from larva to pupa hatching into a chrysalis and then a slow metamorphosis into a butterfly. A variety of specimens were organized beside each other. From behind a glass wall, the head of the nursery walked Rick through the entire process. Rick is not shy when he isn't anxious, and once he starts talking, questions, facts, and stories fire rapidly from his tiny body. You could see him relaxing, becoming preoccupied with the science that had fascinated him in books and was now in front of him. Soon, without my prompting, he asked if he could touch them behind the glass. He did.

And so we returned to the main conservatory with swarming butterflies. He would still cower, toss out mini-shrieks, and hide behind me. Just as I did in my office, I became a model for him, letting them land on me and showing him what was unique about a particular butterfly compared to the others, always reminding him of why they are cool and interesting. And that seemed to work . . . until a butterfly landed on him and he flipped out with arms flailing again. We talked about his concern that people are going to stare at him disapprovingly, thinking he's a loser because of his reaction when butterflies get near him. I said that we should test out this idea. People are paying $15 to enter the conservatory because they want to see butterflies and after a few seconds elapse, people are going to stop looking at a screaming visitor and go back to why they paid their entrance fee. To test the idea, I told him to watch me. I re-enacted what he thought was the doofiest, attention-grabbing thing he does. His job was to watch other people watching me and learn that we often think people pay attention to us much more than they actually do. We think the spotlight is on us when, really, people could care less about us. I ran around the conservatory, shrieking, "oh my god, oh my god, oh my god, butterflies are so scary, so scary, get away, get away, so very scary, the devil's work!" He was laughing hysterically when I came back. I was pleased to point out that while he was watching me, he didn't notice the two butterflies hanging on his shirt.

And then I asked him to join me, and he did and he had a great time. I remember him telling me, "I wonder what people are going to think?" and "I wonder if they are going to kick us out?" The tone was different, as he was really interested to see how people react. More important than the words was his attitude—he was being playful. He was having a good time. All the while, butterflies were landing on us, and lo and behold, no one really cared. They smiled at the spectacle. After a few more trips, Rick was a different person. He spoke intelligently about butterflies, his posture was relaxed, and when we ate lunch, he wasn't just talking about butterflies, he was rapidly switching just as we did in our car rides. A few months later, Rick's mom told me he enrolled in sleep-away camp for the first time and made lots of friends.

When we cranked Rick's Explore knob and brought his interests to the forefront, he did things even when his Anxiety knob cranked up. Soon his Anxiety knob began turning down, but he didn't even notice because he was busy living. When we can clarify our passions, we can get our hands and feet moving toward them even when anxieties enter the picture. Rolling out of the quicksand, with anxiety and self-doubt stuck to our skin and clothes, we enter the larger world without waiting for the day that will never come when we are spotlessly clean.

These and other exercises are designed to open up your eyes to where your life is heading. By being more accepting of anxious thoughts and following your curiosity, you allow yourself to be directed toward more fulfilling moments and a more fulfilling life. All of these strategies are a different way of thinking about anxiety and other unwanted experiences. Instead of trying to get rid of them, we focus our movements toward what is most meaningful, taking the anxiety with us that is guaranteed to rise and fall again and again throughout the journey.

8

The Dark Side of Curiosity
Obsessions, Sensational Thrills, Sex, Death, and Detrimental Gossip

A strange contradiction exists when we closely scrutinize curiosity. Despite all of the benefits of being a curious explorer, there is a tipping point. Extreme curiosity at the wrong time or in the wrong people introduces a dark, destructive side. Just like Russian nested dolls that can be pulled apart to reveal yet another, smaller version inside, the dark side of curiosity intrigues us. To some degree, this is normal and relatively harmless—think of rubbernecking at the scene of a fatal accident, reading newspaper articles about teenage arsonists, and following the exploits of celebrities who are living train wrecks. The more extreme or weirder it gets, the more curious we become.

To date, positive psychology writers and researchers have had little to say about the yin and yang of the positive and negative. Most of the discussion focuses on optimism, gratitude, love, positive emotions, bravery, and happiness, as if more is always better. But a fair and balanced story about curiosity has to acknowledge that too much of any positive experience or "strength" can become a liability. Depending on how it's expressed, curiosity can be perverted, annoying, meddlesome, impairing, and even deadly. Too much curiosity can lead to obsessive behavior, dangerous thrill-seeking, and detrimental nosiness.

A fuzzy boundary exists between healthy and unhealthy interests.

Interests are generally a healthy expression of investing in activities that are personally meaningful and nourishing. One person's interests can be exciting, enviable, or inspiring, or the object of ridicule, disgust, or boredom. At one party, a group is mesmerized by a marine biologist who shares stories of how seahorses are the only creatures where males give birth, how a certain kind of octopus can change colors to communicate with friends, and so on. At another party, the same biologist is mocked for taking a greater interest in the sex lives of sea creatures than his own. Interests can be bizarre or amusing, depending on the context. Although most are relatively harmless, some aren't.

Obsessive Passions

At 38 years old, Steve Wiebe lost his job on the same day he signed the deed to his new family home. Married with two kids under nine, unemployment fueled his self-loathing. Steve wanted to succeed at something, to be the best at something, and this was something that had eluded him his entire life. Of endless possibilities he might have pursued, he homed in on breaking the 25-year-old world record score of a classic video game called Donkey Kong. He bought an arcade machine and kept it in the garage. To pacify his wife's mixed feelings about his new interest, he practiced only at night—that is, every single night, for months.

On one particular day he was making inroads on the record. All of his playtime was videotaped so that he could send it in to the gaming authorities who keep track of these historic events. Steve was alone, presumably watching his two kids, and working his magic with the joystick when his youngest son started bellowing (as captured on the tape):

Dad!!
I can't believe this is happening, guys.
Wipe my butt!

Wipe your bottom? I will in a second, bud.

Can you bring me some toilet paper?

Derek, Derek, I'm going to get the world record.

Don't play! Don't play!

Derek

No! No!

Derek

Stop playing . . . Donkey Kooooooong!

Derek. You stop it.

And so the backtrack on the tape went as Steve sat mesmerized in front of the screen for hours while his kids called for his attention. Steve ended up crushing the world record that day. On the car ride to getting his Donkey Kong score immortalized in the *Guinness Book of World Records*, he explained to his nine-year-old daughter, Jillian, what it meant to him. She wisely replied, "Some people ruin their lives to be in it."

For many, this sort of obsessive passion comes at a price of broken marriages, deteriorating health, and disconnect from the rest of society. Steve was fortunate that his wife recognized how important it was for his self-esteem to conquer the demons of a lifetime of failed conquests.

Many of our behaviors are motivated by the attainment of some kind of reward, whether extrinsic or intrinsic. Extrinsic rewards are derived from our environment such as grades, money, and praise from others. We work hard or take on a task to get a carrot. Intrinsic rewards, on the other hand, are things we do for their own sake, regardless of the external payoff. Examples of intrinsic rewards include personal fulfillment, satisfying an interest, and experiencing a sense of pleasure or accomplishment.

Athletes and musicians are excellent examples of curious explorers motivated by intrinsic rewards. Consider the ballet dancer who practices daily despite painfully blistered feet. Extreme athletes run ultramarathons, attempt to swim the world's oceans, and peak Mount Everest. A boxer finds the stamina to prevail until the final round, nose broken, lip bleeding, and vision fading. Sure, these performers may also be motivated by the cheering crowd, lucrative contracts, and recognition. However, the hours and effort athletes and artists devote to their pursuits and the pain they endure are typically about more than the elusive glory of obtaining a trophy or an encore. To sustain these long-term interests and to excel, these people are motivated by something more—passion for what they do. Researchers have found that athletes and musicians motivated solely by extrinsic rewards perform worse than their abilities allow, get injured more often, have worse relationships with peers and coaches, and quit more easily during difficult times.

Not every activity that sparks an interest becomes a passion. Of your various hobbies and interests, passions arise when you enjoy the investment. Passionate activities begin to define a person and become a fundamental source of meaning.

Consider 13-year-old Kyle Krichbaum. Whereas most children ask their parents for giant water guns, comic books, or bicycles, Kyle asked for a vacuum. According to his parents, Kyle has been fascinated with vacuums since he was an infant—he was mesmerized when his mother vacuumed the house. So his parents gave him his first toy vacuum at the age of one. At the age of two, Kyle dressed up as a Dirt Devil for Halloween. Soon, Kyle began to incorporate vacuums into every aspect of his life. At school, his artwork had vacuum themes such as (the first?) vacuum calendar. At home, he created vacuums out of cardboard and Lego pieces. By the time he was in elementary school, Kyle was vacuuming his classroom and principal's office during recess. It's not that Kyle didn't like playing with other children, he preferred to vacuum.

Today, at age 13 Kyle owns more than 165 vacuums and uses almost all of them—in rotation. When Kyle is not in school, he can be found vacuuming his house, up to five times each day. Kyle's extensive collection includes the first vacuum ever made in 1907, the Hoover Model O, now worth an estimated $10,000. In Kyle's house, vacuums can be found in every room of the house, with the exception of his sister's room.

Intrigued by Kyle's passion for vacuums, the media highlighted his story in newspapers and talk shows. He was even invited to speak at the

2008 Vacuum Dealers Trade Association national conference. A vacuum savant, he found a way to turn his passion into a business by repairing vacuums for neighbors. With his profit, Kyle buys more vacuums. This is all a great experience for Kyle, who wants to eventually own a vacuum repair shop. But what Kyle really wants to do is create the first soundless vacuum.

But setting these positive benefits aside and acknowledging that Kyle's parents don't mind having the cleanest floors in the neighborhood, they would prefer he go out and play more often. He is at a crucial age for learning effective social skills. If over time Kyle continues to spend time alone with his vacuums, it may become difficult for him to build successful relationships. In addition, we can see how Kyle's vacuuming disturbs his family life. For Kyle's parents, they must be careful never to leave anything of value lying on the floor. And Kyle's sister wishes for a few moments without the endless buzz of a vacuum.

Canadian psychologist Robert Vallarand has identified two types of passion—harmonious and obsessive. Harmonious passion refers to activities that we choose to engage in on our own terms. We decide when to participate and when to stop. With this flexibility, we are able to reduce our involvement or stop anytime it becomes a hassle. Many if not most of us are able to find passions that occupy significant spaces in our lives while leaving room for other pursuits that we care about.

People with obsessive interests feel controlled by their passions and experience unwanted pressure to persist even when their minds and bodies say otherwise. This pressure might come from the outside world such as parents, coaches, and teachers, or it might come from internal motives such as guilt or a need to be accepted. Vacations, holidays, sicknesses, and other disruptions (positive or negative) to their obedient devotion are frustrating. Curious exploration is stifled as they ruminate about their lack of dedication and what they should be doing instead of being mindful, appreciating whatever lies before them in the present. Trapped in the past or future, their well-being is compromised.

Obsessive passions are all-consuming and people become dependent on them in a similar vein as alcohol and drug addictions. While this persistence may lead to benefits such as commitment and skill mastery, the lack of flexibility sucks people away from life's other offerings.

Obsessive passions also negatively impact physical health. While most surfers enjoy the ocean in the spring, summer, and fall, it can become dangerous in harsh winter or storm conditions. This knowledge does nothing to deter obsessive surfers. The same goes for athletes with injuries and writers with headaches. People with obsessive passions ignore signs of deterioration, and by continuing with their pursuits, their maladies only worsen.

At this point, it's important to differentiate obsessive passions from the obsessive acts found in people suffering from the psychiatric problem known as obsessive-compulsive disorder (OCD). People with OCD engage in repeated compulsive behaviors (checking to see if the stove is off, counting stairs) in hopes of neutralizing their unwanted thoughts and impulses (such as images of nude children frolicking in a playground). Normally in OCD, there is no curiosity and no intrigue about these uncontrollable thoughts and behaviors. In fact, they are quite distressing, which explains their attempt to avoid them. People with OCD are curious about their passions, even if they don't always enjoy them.

For both Steve and Kyle, strong interests degraded into obsessive passions. When engaging in activities we're passionate about, it's easy to lose control and remain oblivious of the consequences. Curiosity-guided activities have the potential of becoming obsessive passions. The content of obsessive passions tends to be relatively benign. Sometimes the obsession ends up being another human being.

Unrequited Love and Erotomania: Undying Curiosity for a Disinterested Object of Affection

Earlier, we talked about the universal desire to grow as human beings and the important role other people play in that process. We are attracted to people who help us to expand who we are and how we see ourselves. Unfortunately, our feelings of attraction aren't always reciprocated.

When it comes to unrequited love, the desire for a person is enormous but the probability of actually being in a relationship is near zero. For myself, I can enjoy my crush on Jennifer Connelly (from the movies *A Beautiful Mind* and *Requiem for a Dream*) knowing the only attention I might receive from her is a stock photo with a stamped signature—or a pistol whipping from a body guard if I ever get close enough to say "hi."

Unrequited love becomes much creepier and perverted when people ignore outright rejection and stalk the object of their affection. They believe a relationship is possible if they just try harder. They may believe that their love interest is also in love with them, or, that if they ever met, the object of their affection would instantly feel the same way. Curiosity goes awry. Their intrigue about what their love interest is doing, and how they "secretly" show their feelings for them, adds an elusiveness that makes it even more exciting. The uncertainty and excitement of the chase gets mixed into the love cocktail, fueling further interest and persistence.

Due to the public nature of their lives, celebrities are common targets of obsessive passions. The name for this kind of romantic interest in someone with a higher social status (e.g., politician, celebrity) is *erotomania*.

John Hinckley, Jr., attempted to assassinate President Ronald Reagan to gain the love and respect of actress Jodie Foster. Margaret Mary Ray is known for stalking and claiming to be the wife of late-night-TV host David Letterman and former-astronaut Story Musgrave. Other celebrities known to have erotomanic fans include John Lennon, Linda Ronstadt, Madonna, Steven Spielberg, Barbara Mandrell, and Uma Thurman.

Dr. Park Dietz analyzed more than 5,000 letters written to celebrities from delusional fans. Dietz found that at least 800 of the fans had made an attempt to approach the celebrity in person. When the fan failed to get a response, they often became more persistent and increasingly violent.

The dark side is not limited to obsessive passions. Sensation seeking, morbid curiosity, and eccentric sexual interests serve as themes in a subset of peoples' lives. People with these interests make up a relatively small percentage of the population, yet in our voyeuristic society, we are intrigued by them. How did they develop these extreme forms of curiosity that violate social etiquette and, in some cases, the law? What causes someone to put their own or someone else's life on the line for the sake of an interesting experience?

Social Information Seeking

Curious people can also be downright annoying. Does this characterize you?

- My friends would describe me as always peppering them with questions about things.
- I sometimes irritate friends and relatives by asking them very personal questions about why everything is the way it is.
- People tell me that I am often so distracted by new, exciting things that cross my path that I end up being tough to be around.

It's fair to say that in the absence of truly caring about people and wanting to know them better to connect with them, being curious is a social liability. People know when they are being treated as if they are interchangeable with other forms of novelty. Nobody wants to be exploited as a vessel of information until someone more interesting strolls along.

Gathering information about people in our social world helps us navigate our social environments. The type of information that we seek about others can be public (asking about someone's profession or gender) or private (wanting to know who someone is sleeping with). In most cases, obtaining information about other people is acceptable and even expected. This is because social curiosity serves several important functions:

- It feeds our curiosity and increases our knowledge about others.
- It provides us with a sense of belonging.
- It helps us evaluate ourselves and our social standing compared to our peers.
- It has entertainment value.

The motivation to seek information about others varies. Often, we respect social curiosity. We expect people to verify the credentials of a nanny or a surgeon and learn about the family and friends of romantic partners.

Many people find it intriguing to gather information about people who are far removed from their lives. Our culture's fascination with celebrities is a great example of social curiosity. Supermarket checkout racks are filled to capacity with tabloids that feed our interest in the lives of celebrities and noncelebrities alike. Is this social curiosity problematic? Hardly. Yet like any personal interest, it's possible to get carried away.

Social curiosity can lead to the violation of privacy through spying, snooping, and eavesdropping. Of course, listening in on another's conversation, reading a sibling's diary, or checking a child's backpack to ensure that homework is being finished are relatively harmless acts. The consequences of getting caught (for instance, a heated argument), are unlikely to be severe or long-lasting. But what happens to the employee who intentionally accesses classified company information? The neighbor who sifts through other people's trash? The secretary who peruses the medical charts of physically attractive clients to whom she has no legal access? In each of these examples, the desire for information is disturbing, prohibited, or illegal.

Using covert tactics to access information can damage personal relationships. The privacy-violating acts of people with extreme curiosity may lead to rejection, making it increasingly difficult to satisfy their curiosity. When the information acquired is shared with others, as it often is, the repercussions can be even more severe.

Social curiosity has a complementary behavior: gossip. Gossip is pervasive and we begin soon after we learn to speak. Some researchers have found that as much as 60% of adult conversation can be classified as gossip.

Gossip is pervasive because it's useful. We can't be everywhere, and gossip lets us learn about the world, predict and understand other people, develop relationships, and learn how to adapt to new situations and groups. We use gossip to entertain, gain favor and create tight bonds (trading confidential secrets), and avoid awkward silences. Gossip allows us to learn and teach others what behaviors are acceptable and what's frowned upon. Gossip about the man who slept with his married co-worker allows others in his community to better judge who's trustworthy, how much a person can get away with, and what alliances are being formed.

Try to go cold turkey and avoid gossip for a single day and you will discover how much it infiltrates your conversations and consciousness. Gossip often involves judgments about the third party's character, reputation, or social status. Among the most common topics are sex, money, illegal or salacious activity, and anything considered abnormal.

Simply sharing facts about another person doesn't constitute gossip ("Karen had the baby!"). This statement becomes gossip if you speculate or evaluate her actions. ("I don't think she's going back to work," "there is no way she can survive as a stay-at-home mom!") For gossip to work successfully, the information being spread needs to be viewed as truthful, clear, and comprehensive. When the information is ambiguous, unverified, or incomplete, gossip often wanes.

If gossip serves so many useful purposes, why must we feel guilty? Why do we gossip in hushed tones? There's a tipping point when gossip ceases to be useful and instead harms.

Gossipmongers are typically highly curious people who invest far too much time and energy on the lives of others. Frequent gossipers tend to be popular, but also less likeable. Their high level of popularity makes sense. People want access to interesting stories. As the gatekeepers, breaking confidences in their social circle, they also end up as targets of anger, scorn, and, yes, gossip.

We can use gossip to improve our social position at the expense of others. In fact, there is a greater likelihood you will spread malicious gossip about another person when they are viewed as a rival.

What mitigates against spreading malicious gossip is the fear of being named as the source. Opportunities to be anonymous bring out the worst. One gossip website aimed at college students prides itself for allowing anonymous postings. Topics include "outing gay men who have yet to 'come out'" and posting racial slurs. A web post that asked "Who is the sluttiest girl at the University of _____?" received more than 300 replies within days. As the targets attempt to repair their damaged reputations, stories of despair, loneliness, and suicidal gestures are far too common.

Sensation Seeking: The Need for Speed, Thrills, and Excitement at Any Cost

To fully understand curiosity, we have to understand people who crave the most intense novelty and thrills they can find. These people are best described as high-sensation seekers. Love or hate them, sensation seekers are rarely ignored. Consider two high-sensation seekers immortalized in history.

Person A

He grew up in a relatively poor family. Buried in his books, he saw college as his ticket to a better life. It paid off. Earning an academic scholarship, he attended a small community college. He excelled and was offered a scholarship to study alongside some of the best students in the world at Stanford University. In this new terrain, he was transformed. He was good looking, stylish, witty, and refined in his manners. He worked as a counselor for a suicide-crisis hotline, saved a three-year-old boy from drowning in a lake, and the local police department called him a hero for apprehending a purse snatcher on the street. After graduating, he became a rising star in the Republican Party and worked on the committees of various governors and congressmen. When he was accepted into law school, the opportunity for him to enter the spotlight finally arose.

Person B

He needed to understand how things worked and so he set up bizarre experiments. No living creature was spared. Waiting until mother hens left their nest, he would steal fetuses from eggs. What other families kept as pets, he dissected. It's unclear just how many rabbits, birds, cats, and dogs took their last breaths in his hands. Many of his nights were spent robbing graves. He would unearth bodies buried by loved ones only hours before. With awe and wonder, he peeled back flesh, cut through tissue and organs, and removed the smallest shards of human remains. He covered his tracks well. His actions led to public outcry and fear in the local news. Dissatisfied with the pace of expanding his "collection," he turned inward, experimenting on himself. Among other acts, he stabbed his own penis, regularly scrutinizing these self-inflicted wounds.

That well-educated, politically ambitious man who overcame childhood poverty is Ted Bundy, perhaps better known as the most notorious serial killer in American history. He confessed to raping and murdering at least 20 women, and it is believed that his murder spree spanned nearly half the fifty states and claimed as many as 100 victims. In 1989, he was executed in an electric chair for his heinous crimes.

That body snatcher is none other than John Hunter, considered by many to be the father of modern surgery. Despite his towering scientific achievements, scarcely anyone outside of the medical establishment knows his name. The reason is that his brother and other scientists continually stole or plagiarized his work. John Hunter avoided confronting his brother because he cared more about satisfying his curiosity (intrinsic rewards) and less about acknowledgment and prestige (extrinsic rewards). It's a shame because if he was given credit for everything he accomplished, his name would be in the ranks of Issac Newton, Albert Einstein, Nikolai Tesla, Marie Curie, and other titans of scientific advancement.

Countless human lives continue to be saved because of his ground-breaking eighteenth-century experiments. Around 1850, in Britain, the life expectancy of the average person was pathetic, only 39.5 years! Deplorable public health conditions and beliefs in magic and superstition led to unnecessary deaths at the hands of incompetent physicians. John Hunter changed this mentality. Hunter's intense curiosity and self-confidence led him to establish new guidelines for training doctors and treating patients. He believed that medical procedures needed to be tested to see if they were effective. He believed doctors needed to practice the actual procedures they were going to perform on patients. Before Hunter, surgeons harnessed their skills from reading books, only reading books (think about this the next time you complain about your physician). As a personal sacrifice in his quest for knowledge, he injected himself with syphilis (stabbing his penis) to better understand venereal diseases. Albeit controversial and gruesome, Hunter's experiments provided the knowledge for veterinarians, physicians, and scientists to do less harm. Hunter educated the next generation in an endless attempt to discover the nuances of how life operates.

Sensation seekers are people willing to take on physical, social, financial, and legal risks to satisfy their need for novel and intense experiences *for their own sake*. They despise being bored. Often, risk is necessary to satisfy their desire to feel alive. But risk-taking is a means to an end, with the rewards being new, exciting experiences.

Low-sensation seekers aren't merely adverse to risks. They see no point and can't find a good justification for seeking out thrills. Low-sensation seekers are more sensitive to pain and other stimulation. They don't value new sensations. This is why high- and low-sensation seekers can't understand each other and make bad bedfellows.

Being a high-sensation seeker isn't necessarily good or bad. It depends on how intense desires for novelty and variety are channeled.

Sensation-seeking includes taking a walk in a cold breeze, meditation and yoga, and the thrill of talking to new people. It affects preferences for food, music, movies, and conversational topics. What if you were invited to taste unusual foods like snake, slug, gecko, puffin, reindeer, or whale sashimi? For the uninitiated, a few squeamish calls for mommy are natural. This is the province of the high-sensation seeker. They are less likely to let their initial disgust deter them from a new experience.

High-sensation seekers also enjoy less socially appropriate interests such as gambling, cliff diving, drug use, aggressive driving, promiscuity, and for a few, violence and serious criminal activity. High-sensation seeking isn't a disease or a recipe for deviancy. It's a personality trait. Some high-sensation seekers happen to be evil incarnate, but others are successful creators and leaders. Being open to new experiences and pursuing ideas that rival conventional wisdom is essential to society. If people weren't willing to take risks, our species would stagnate. The acts of a few high-sensation seekers ensure that we continue to evolve.

Dangers behind the Wheel

You don't want to be driving on the road surrounded by a pack of high-sensation seekers. You don't want to loan your car to a high-sensation seeker. Sensation seeking is related to nearly every type of risky driving behavior: driving while intoxicated, driving more than 80 miles per hour, not wearing seat belts, keeping unsafe distances from other drivers, swearing at other drivers, bobbing and weaving in traffic, racing other cars, using the horn when annoyed, and giving people the finger among other rude gestures. These drivers are more likely to cross solid lane crossings, receive moving violation tickets, and trigger car accidents.

High-sensation seekers are dangerous because risky driving is exciting and thrilling to them.

Risk-Taking and Intimacy

The high-sensation seeker is at greater risk for nearly every type of risky sexual behavior: sex at a younger age, high number of sexual partners, sex with multiple partners, unprotected sex, sex with strangers, sex with prostitutes, combining drugs and sex, and unwanted pregnancies. Being a high-sensation seeker doesn't mean these sexual experiences are going to be a problem, but the potential danger is substantial. For instance, high-sensation seekers are more likely to have sex with someone with genital herpes, HIV, or AIDS. To them, the short-term pleasures often outweigh the costs. Worse, they often minimize the costs, being less likely to use condoms and birth control. Thus, there is an overabundance of HIV and AIDS patients who are high-sensation seekers.

The sexual risk-taking linked to sensation-seeking is just one contributor to their difficulties in forming lasting, meaningful romantic relationships. High-sensation seekers tend to see love as a game of flirting, arguing, and challenging each other. This also includes devaluing faithfulness and commitment, with sexual mischief the norm rather than exception.

Like highly curious people, high-sensation seekers recognize the value of spending time with people different from themselves, people who complement and expand them. They also don't view conflict in black-and-white terms. Showing distress tolerance, arguing is often a vehicle for enhancing relationships—a source of sharing, brutal honesty, excitement, and energy. High-sensation seekers open the door to greater closeness and intimacy, and games and play have their place. But treating love as something to be the winner at (you give me more than I give you) is toxic for creating healthy relationships.

Having a partner in constant worry that they are going to bore you doesn't help a relationship. Supporting this idea, high-sensation seekers are less satisfied in their relationships and more likely to have affairs and divorce their mates than anybody else. Even worse things happen when you put a high- and low-sensation seeker together. Infidelity, violent arguments, stonewalling, and breakups are extremely likely. They just don't understand each other's interests.

There is one exception to these relationship problems. Partners with similar high-sensation-seeking tendencies tend to be extremely compatible, constantly finding ways to stimulate and challenge the other.

The Appeal of Criminal Activity

The search for thrills and excitement can also lead to illegal, criminal behavior that is far worse than speeding on the highway and serial one-night stands. Why do people break into homes but take nothing from them? What would motivate a highly-paid and well-known celebrity like Winona Rider to shoplift from Saks Fifth Avenue? Within each of us lies a certain intrigue with deviant behavior and breaking the rules. For those who live seemingly straight-laced lives, this curiosity lies dormant. It's so weak that it can be pushed out of consideration by rational thought. Then at unexpected times, this intrigue awakens, motivating us to explore dark crevices we know we shouldn't enter. It's the desire to know what it feels like to commit crime. It's the need to experience the uncertainty of getting caught. I interviewed a highly intelligent, well-mannered woman in a white-collar job who shared her delinquent moments.

I have been privileged enough to never lack any material possession or be of any distance from monetary access. But at a time when I was working in Europe, I found myself shoplifting small-cost products from street kiosks. I recall thinking that street kiosks practically invited theft. One vendor, products strewn about, many hiding spots, and easy escape. So one day I tried it. I swiped a piece of 5-cent gum. To my surprise, I didn't feel any better or worse about myself afterward. I was intrigued at how easy it was. Soon I began stealing bubble gum nearly every time I went, regardless of whether I was buying anything else. Then on a couple of occasions, I decided to up the risk and go for something larger—like a water or juice bottle. Again, it was so easy. I was content knowing that I could do this. I didn't feel ashamed. My crimes were petty. I wasn't ruining someone's livelihood. Sure, later I upped the stakes a few times, but I wasn't doing major harm. I loved the excitement of outwitting the system. Just a few times, here and there. Holding my stash, looking at it later, I felt seductive.

The exhilaration of not getting caught is a part of these illicit, sensation-seeking tendencies. Criminal acts add a pleasurable dimension to their personae.

At the far end of the criminal spectrum are murderers and rapists like Ted Bundy and the Ku Klux Klan. Originally, the KKK was designed to be an innocent escape from boredom. As historian Wyn Craig Wade describes, "A few charismatic leaders lured in unhappy, uneducated youth looking for thrills and adventure. Wearing bedsheets to look like ghosts, they had fun scaring slaves who feared ghosts and demons and held superstitious beliefs about the dead. Taking advantage of their frustrations and failures to gain social mobility, the leaders of this sect slowly indoctrinated their members into seeing how superior they were, how they could scare and terrorize anyone. The playful pranks degenerated into torture, castration, tarring and feathering, beatings, and lynchings. Violent thrills originated from humble beginnings.

Hannah Arendt reached this conclusion when she wrote of "the banality of evil." Meditating on Adolph Eichmann and the Nazi regime, in a moment of clarity she realized that great evils in the world are committed by ordinary people. Seduced by the appeal of harming other people and ruining their lives, violence often originates as an outlandish idea for finding fun and excitement. Burning forests, robbing banks, raping men and women, and setting off bombs can be an antidote to everyday tedium and pain.

This idea doesn't sit well with us. It's easier to envision acts of senseless violence being committed by monsters who are nothing like us. Although it is often true, a person with high-sensation seeking who happens to also be impulsive, aggressive, and lacking in self-control can find themselves being violent to entertain themselves. Surrounded by deviant peers and without proper guidance, curiosity can lead them to "the intrinsic appeal of evil." Of course, we are talking about extremes. Saints and sinners are the endpoints on the continuum of curious people.

Sensation Seeking at Work

High-sensation seekers try to find work that ensures intense novelty and uncertainty on a daily basis (or they suffer). When you look at the sensation-seeking levels of firefighters and police officers, they are no different than the average criminal. Scientists also found that the sensation-seeking levels of criminals are the same as people devoted to extreme sports such as snowboarding, skydiving, and rock climbing. It makes sense. These outlets provide similar thrills and excitement.

Even in the throes of major combat, war veterans report excitement and meaning. Prisoners of war report thrills and excitement along with their terror and anger. Mixed feelings are common, particularly for the high-sensation seeker. They accomplish great, heroic feats by taking risks and working with instead of against their fear. They cope best in difficult, life-threatening situations.

Combat veterans who are high-sensation seekers report less trauma, stress, and emotional problems in the war zone. However, they often have greater difficulty making the transition from the excitement of wartime activity to the domesticated life of commuting to work, taking out the garbage, and trimming hedges. The mundane, quiet routine is a bad match for their need for speed and thrills. Knowing this, we can improve the lives of servicemen and -women by helping them find new, healthy outlets for these needs. Ignoring their biologically based desire for variety and novelty is a recipe for unnecessary pain and suffering.

Other high-sensation seekers become psychologists, surgeons, trial lawyers, and investors. In contrast, low-sensation scorers prefer structured, well-defined occupations with order and routine. However, sometimes it's less about what people do for a living than how they approach what they do. High-sensation-seeking parents find rewards that low-sensation seekers may not be attuned to. For mothers, this might include a more intense sensory experience of breast feeding, relishing the spontaneity of their older kids, and letting their children's playful and creative games influence them to do the same. Although parents are the guardians and caregivers, our kids can be great teachers to us if we are open and curious.

In one study, parents and teachers were asked what they learn (if anything) from children. Life lessons included:

- Shifting life priorities. Stop dwelling on minor hassles and redirect energies toward meaningful life aims.
- Being more creative and flexible. Realizing there are multiple ways to reach a given goal (and get out of a jam).
- Looking at the world with greater wonder and awe. When children ask "why," we are reminded how rarely we question what goes on in the outside world. We know much less than we think. Kids provide a model of how to be ignorant without any semblance of embarrassment. The pleasures of finding things out outweighs the discomfort of not knowing something.
- Appreciating the pleasures of being playful and goofy. Far too many adults mistakenly believe this should end at a certain age and be replaced with a serious, somber attitude. Children remind us of what we're missing without any good reason.

High-sensation seekers are more apt to recognize these life lessons and be more open to adopting them. By finding sensations and transforming situations to be more interesting, it becomes easy for sensation seekers to get the most from their occupations.

In general, high-sensation seekers get bored easily, and when they do, they initiate problems, including arguing with coworkers or stealing office supplies to fuel excitement in an otherwise underwhelming environment. Having to deal with volatile, spontaneous high-sensation-seeking employees, managers and coworkers often suffer. The best way to deal with the situation is to ensure there is a good match between a person's personality and the tasks they are given. Variety is essential. Asking and giving feedback is essential. In the end, if a sufficiently stimulating environment can't be provided, sometimes the high-sensation seeker needs to move on to a high-throttle, exciting career.

Channeling Sensation Seeking

We can see signs of sensation-seeking in children as young as two years old. Toddlers who grow up to be high-sensation-seeking adults are:

- Quicker to reach for and react to new toys and sounds
- Choosing more intense and stimulating toys
- More likely to explore
- More likely to take physical risks
- Less distressed when their parent leaves them when there is sufficient entertainment

Finding a good way to channel their curiosity and energy is the key from early childhood all the way into adulthood. Social influences—from parents and teachers to friends, coaches, and mentors—play a huge role in what happens. An adolescent sensation seeker can be attracted to risky sports and taken under the wing of a caring coach. He might just as easily be attracted to crime and drugs and taken under the wing of a local drug dealer. High-sensation-seeking teens are about three times more likely than low-sensation-seeking teens to experiment with illicit drugs.

Mentors and peers are important. High-sensation seekers raised in positive and loving homes are less likely to engage in negative and illegal thrill-seeking activities. The war on drugs, violence, and crime would benefit from a closer look at the world of curiosity. It is one of the primary motivators for why people do things, and given inadequate guidance, it's easy for people to take a turn toward the darker side.

Teddy Roosevelt, Chuck Yeager, General Patton were all high-sensation seekers. Some historians describe each of them as antisocial thrill seekers who were able to lead successful lives (to say the least). It's all about finding the right environment and mentors at as early an age as possible. High-sensation seekers can be guided toward healthy substitutes that satisfy their need for intense novel and varied experiences.

Eccentric Sexual Interests

When I was a teenager, my uncle had a bizarre store in New York City. In the front, you could find a slew of goods to rummage through, including old beta VCRs with play buttons the size of a baby's torso, used CDs, switch-blade knives, and graduated cylinders from defunct chemistry labs. In the back room, it was another story. Open the curtains and you were welcomed by a wall of dildos and towering stacks of pornographic videos and magazines that reached halfway to the ceiling. My twin brother and I worked at this metropolis of lust at far too young an age. At the counter, we told customers about the fantastic deals they could get— buy twelve porno videos and get the thirteenth free! With little more than peach fuzz below our noses, customers would ask for movies by title, each more explicit than the last. Needless to say, I developed normal sexual interests rather quickly and at the same time became much more interested in the eccentric extremes.

Before heading off to college I took a few bizarre films home with me. I brought these films to college and found great interest in people's various reactions. Some people couldn't stop laughing, others would clearly be aroused, some were aghast, and then there was an acquaintance who sat there frozen, eyes wide open, and without warning, leaned down and vomited all over himself (he went on to become a weatherman on a major network, so I won't divulge his name). I am not defending pornography and certainly wonder what would have happened to my personality if I spent my young adult life watching *Star Wars* and *Jaws* like everyone else. What is interesting is that some people are drawn to the lurid and morbid because it's what gratifies them. Some people understand it, some people don't. We need to investigate these interests because they provide a lucid understanding of where some curious explorers travel.

When deciding how to spend their free time, we turn to the interests that invigorate. In most cases, these activities are rather innocuous. As indicated throughout this book, reaping the benefits from curiosity is best realized when the outlet is welcoming and not intimidating. Yet for some, these outlets can be dangerous or cross the boundaries of what we as a society deem appropriate.

Sexual Obsessions

Think for a moment about what gets you sexually aroused. Seeing your partner dressed in a certain way? A particular scent? Watching an erotic film? Sexual preferences and the sources of sexual arousal are highly individualized. What is arousing to you may be abhorrent to your neighbor. Sexual arousal can be elicited by a vast range of things, from common to uncommon parts of human anatomy (breasts, buttocks, thighs, ears, clearly defined clavicles) to nonhuman objects (animals, garter belts, lingerie, shoes), and the grotesque (urine, feces, amputees, dead bodies). If you are cringing at this moment, you are not alone. Whereas most perverse thoughts, fantasies, and sexual practices are harmless, some people (or their victims) experience significant distress regarding their sexual experiences.

People with paraphilia possess sexual urges and fantasies that revolve around something other than a full-bodied adult. Their sexual interests often become obsessions, in which enormous amounts of time and energy are exerted to satisfy lustful cravings. Sexual desire might arise from experiencing pain and humiliation at the hands of another (masochism) or doing it to someone else (sadism). It might arise from exposing one's genitals to a surprised audience (exhibitionism), spying on others, usually while they are disrobing or having sex (voyeurism), or touching and rubbing one's genitals against nonconsenting others (frotteurism). It also includes being sexually attracted to inanimate objects or isolated body parts (fetishism). More disturbing is when adults focus their sexual fantasies and behavior toward prepubescent children (pedophilia). More commonly seen in men, paraphilias tend to co-occur with other emotional problems, especially substance abuse, depression, and anxiety disorders.

You may also be wondering how people can come to have such seemingly bizarre sexual preferences. Just as their sexual preferences are diverse, so are the reasons. Focusing on the desire for novelty, uncertainty, and surprise is just one of many influences in the development of unusual sexual behaviors.

Suppose you happen to crave sucking on calloused toes, and this is the only way you can quench your sexual thirst, and you find an open and willing romantic partner. A bona fide fetish exists. You are not harming another person or yourself, and you are doing it all in private. Sure, it's weird and interesting—so is dressing up with your partner as jail warden and prisoner, librarian and eager student, or Gargamel and Smurfette.

To further illustrate the complexity of when to label these atypical sexual interests as "disorders" or "perversions," consider the sexual need to observe others. It's common to experience some level of nonsexual or sexual satisfaction from observing other people. At some point in an average week, millions of Americans will enjoy watching privacy-violating reality television programs. Others will wade through pornographic material.

What separates the millions of Americans who become sexually gratified by watching others from those known as voyeurs is whether the target being observed consented. In most television and Internet documentations, the actors and models willingly placed themselves in situations knowing they could be observed by the general public. Voyeurs derive pleasure from finding nonconsenting targets.

When you move away from sadomasochism and fetishism, when you move away from consenting partners, it becomes less controversial to call them perversions and disorders. It also becomes easier to understand the disgust, revulsion, and righteous indignation it causes others. More commonly known as the Peeping Tom, voyeurism is about invading another person's territory to get aroused. Voyeurs experience sexual arousal by spying on their targets, hoping to catch a glimpse of them undressing, naked, or engaging in sexual acts. While they watch, they often masturbate. Elements of the unknown are everywhere.

Every spying episode could lead to a new target, and there is no telling what will happen while hiding outside a bathroom window. Will they see anything and if so, when? How will their target respond if they notice? Will they invite them in for some sexual adventures? The antici-patory pleasure can be intense. Many voyeurs enjoy the act of spying for the pleasures of uncertainty.

Whether they will be seen or caught adds an extra layer of excite-ment, further increasing their sexual arousal. It's a case of that optimal state of anxiety, the butterflies.

Just as voyeurism has a clear victim, the same goes for exhibitionism. There's nothing wrong with prancing around naked in your own home or wherever it is socially accepted to do so (nude beaches or spas). When a person exposes their genitals to nonconsenting passersby, however, a prob-lem arises. Exhibitionism is the sexual arousal and gratification experi-enced when showing genitals to strangers. For the exhibitionist, flashing themselves is only arousing when the target, typically a woman, is unknow-ing and uncooperative. What the exhibitionist desires is for his target to respond with her own sexual arousal. He doesn't desire a real-world sexual relationship. He wants the sight of his flesh to delight her.

Having this ideal exchange is not likely, of course, and is virtually unheard of. But the slight chance that his victim may respond in this way and with a unique reaction sustains the excitement. Throw in the legal risk of public exposure, and the exhibitionist, like the voyeur, finds plea-sure in the "what-if" questions that linger.

To be clear, it might be enjoyable for the voyeur or exhibitionist, but it is violating the life space of the nonconsenting target. It's not just a ru-ined moment. It can shatter a person's beliefs about a safe and just world. It can make them terrified of other people. It can make them afraid of be-ing in their own home or anyplace where they are alone. When real vic-tims are involved and their lives are altered, and then the dark side of curiosity becomes apparent.

There is no single explanation to account for how and why deviant sexual preferences develop. We do know that some people engage in atypical and deviant sexual practices because of the associated risk. High-risk situations provide a sense of novelty and danger, and the uncertainty is sexually arousing. It's not always the sexual act that causes the excite-ment. It can be the thoughts, fantasies, and anticipation leading up to the act that elicit the greatest pleasure. When it comes to anticipating and seeking pleasure, this is when the dopamine circuits of the brain linked to curiosity go wild. The curious explorer is too gratified during the hunt to be satisfied.

Morbid Curiosity

Do you watch real-life emergency room shows where people are seriously injured and hanging on for their lives? Is it hard to avert your eyes from a gruesome scene? Do you ever laugh at violent scenes or situations? Did you watch the Saddam Hussein execution video? How about the fall of the World Trade Center buildings? How many times?

Death is largely a mystery. Although we are able to study the physical occurrence of death and its aftermath, we have few insights. We want to know about the unknown. In fact many people find it intrinsically rewarding to study and learn about violence and death. What does a knife wound look like? How does the body decompose after? How is it different from one person to the next? How do people behave in the final moments prior to death? How do attitudes toward death differ in other cultures than my own?

Numerous books, magazines, and websites have popped up to address these interests. One website, Rotten.com, provides instant gratification to the morbidly curious. Receiving roughly 15 million hits per day, the website owners claim to document only real instances of bad things happening to people. Interested individuals can access documentation and grotesque images of injury, death, and dismemberment. Those with morbid interests can spend hours amusing themselves by watching YouTube videos of morbid events (e.g., clips of wild animals, fish, and reptiles eating humans). Even the news media is aware that the ratings are much better for bad news than good news. Book publishers and movie producers earn billions of dollars annually by focusing on death, destruction, and chaos.

Most people with a small interest in morbid things can comfortably put their interests on hold and return to a more peaceful reality. Far fewer show uncontrollable urges. Even fewer are devoid of empathy or remorse. Thus, nothing prevents them from more daring attempts to act on these interests. These people may spend the greater part of their days seeking death-related information and events. You may find yourself wondering how people can be this curious about violent and morbid things. We don't have to dig far into the recesses of history to find several infamous cases: Jeffrey Dahmer, Issei Sagawa, and so on. Although their stories are grotesque and likely to turn a few stomachs, they offer insight into what humans are capable of doing. Books and movies about them continue to be financially lucrative because people are intrigued. Morbid curiosity extends beyond the doer to captivated fans. The novelty, unusualness, and mysteriousness of these events are often repulsive and captivating at the same time.

Curiosity has a dark side. To benefit from curiosity without being overtaken by obsessions or pulled into behavior that jeopardizes us and others, we need to confront the darkness with the same openness and explorer mentality we can learn to bring to everything else.

9 Discovering Meaning and Purpose in Life

People say that we're searching for the meaning of life. I don't think that's it at all. I think that what we're seeking is an experience of being alive . . .

— Joseph Campbell

The sun sliced through the windshield, sealing me in light. I closed my eyes and felt the warmth on my eyelids. Sunlight traveled a long distance to reach this planet; an infinitesimal portion of that sunlight was enough to warm my eyelids. I was moved. That something as insignificant as an eyelid had its place in the workings of the universe, that the cosmic order did not overlook this momentary fact.

— Haruki Murakami

Our genetic code is 98% identical with our chimpanzee cousins. Mammals, rodents, and other creatures feel curious, get anxious, and work through conflicts of whether to hide from or explore their environment. Yet one thing stands out as being uniquely human—the deliberate search for meaning and meaning-making. In this chapter, I want to show you a final testament to the value of curiosity. It arrives when we closely examine how we go about making, finding, and creating meaning, which for some people leads to a purpose in life.

A Desperate Search for Meaning

At the center of my own life story is this quest for meaning. A major event kick-started it. My twin brother and I lost our mom to breast cancer when we were 13 years old. Of the two of us, I was the momma's boy. Much of my time was spent leaning against her, while I listened to music or nibbled on a Good Humor Toasted Almond Crunch Bar on a summer afternoon. My father had virtually walked out of our lives when we were two years old and to this day he has not come back into it. Suddenly, we were left without parents at far too young an age.

My wife and friends regularly talk about their childhoods. They reminisce about family trips to the beach and Wiffle ball games in the street that lasted until nightfall. I, on the other hand, hardly remember anything about my life prior to age 17. When my mom died, most of my memories went with her. Plenty of psychologists will tell you that this reaction is normal. Getting rid of unwanted feelings by concealing and suppressing thoughts keeps the pain away.

Having lost most of the love and meaning in my life, I started looking for replacements to fill the emptiness inside me. I didn't know it at the time, but I was coping by making and searching for meaning. Self-exploration is rarely a straightforward road. I bounced from one source of meaning to another. I became obsessed with my body. In middle school and high school, I found great satisfaction in sculpting my pipe cleaner arms and legs, which dangled from my 126-pound frame. I lived in the gym, grunting and convulsing as I tossed around dumbbells and barbells. By adding 30 pounds of muscle to my frame, I felt strong inside my skin. It was a release; it was rewarding. It was the ultimate art form.

But I wanted something more. In tenth grade, I became obsessed with using my body to become an elite athlete. Since I lacked coordination, balance, good reflexes, or speed, my options were limited. The arcane shot put became my new holy grail. Needless to say, this hypermasculine attempt at physical stardom led me far astray from high school popularity, but it did give me a strong sense of self at a time when most teenagers don't know who they are.

After I hit puberty, at the age of 15, I lost my interest in sports and discovered the opposite sex. By the time I entered college at 18, bodies other than my own were my driving interest. I wish I could tell you about the intellectual rewards of going to an Ivy League college. I would be lying if I did.

Like many people, when college ended, I bounced around from job to job. I dressed in claustrophobic superhero costumes to sell trading cards outside toy stores, worked as the mail boy for a law firm, and modeled men's jackets at Oriental auctions. Even when I started on the road to a real adult life by working on the New York Stock Exchange, my lack of satisfaction gnawed at me. In keeping with the times, my search for meaning became more deliberate and obsessive. I read books on Buddhism, science and the mind, and finally made the leap to turn my source of meaning into a career.

I left my high-paying Wall Street career and took a few unpaid positions to help me turn my true interests into a career. Slowly, my source of meaning shifted once more and I found my "calling" as a scientist, teacher, and therapist devoted to understanding how certain people are resilient and fulfilled regardless of the stuff life throws at them. For as long as I can remember, "intense passion" and "dedication" are the first words people use to describe me. My own search, borne out of suffering, led me to find a guiding direction, or purpose, in life.

Looking for Direction

We live in a time of great uncertainty. As I write this, the United States is in an economic crisis, several wars between countries are long underway, and mass genocides such as what is happening in the Darfur region are relegated to page 13 of the newspaper as readers move on to fresher story lines. A number of us are going to have to pursue paths that we never expected because a lot of bad things are confronting us. When I think about what people need, happiness isn't the first thing to come to mind.

Happiness is a sign that things are going well in your life, but it doesn't say anything about the substance or quality of how you are living. Although many happy people find profound meaning in their lives, plenty of happy people don't. Being happy doesn't provide direction for what to do.

To successfully deal with hardships, take on challenges, and live in the present with a receptive state of mind to appreciate and make sense of the richness of life as it unfolds moment to moment, we need more than happiness. Meaning is about gaining insight into what to do and what not to do when we're faced with life decisions, big and small. Meaning enhances our capacity to work toward a future that is most in sync with our deepest values and interests.

We can say our life is infused with meaning when we:

- Start to comprehend and understand who we are and what strengths are at our disposal
- Grasp how to work well with other people and know what kinds of situations energize us and what kinds of situations deplete us
- Find a compelling mission for how to invest the most valuable resources in our life—time and energy—in the most efficient and effective way (with one foot firmly planted to relish the present moment and one foot in the future, striving to live in a way that is consistent with what we care most deeply about)

The reason we search for meaning is that a life without meaning is an empty one. A life infused with meaning is a life that can provide us with a secure, bedrock foundation.

Unfortunately, there is no blueprint for finding meaning in life. Exercising regularly and eating green leafy foods isn't going to cut it. Thinking positively and being hopeful about the future isn't "the secret." Being grateful for what we have isn't enough. Isolated encounters that evoke joy and wonder aren't enough. Having people in our social world that we care about, can be ourselves with, and can count on for help isn't enough. Possessing all of the ingredients for a chocolate mousse cake doesn't mean that you have one. You need to organize the ingredients in just the right way or else you have little more than a barely edible mess in your kitchen.

The evidence is clear that people who are meaning-makers and able to learn from events, gain valuable insights, and grow as a person experience profound health and well-being. I suspect having the best of both worlds—abundant happiness and meaning in life—increases the amount of health, well-being, and longevity at our disposal.

That being said, happiness is a shaky foundation, whereas meaning provides us direction for where to go and what to do when we are faced with a tyranny of options. How do I choose my friends? How do I choose my romantic partner? How do I choose a career? How do I weather financial-, social-, and health-related crises? What should I regularly commit myself to in my free time? What made me who I am, and what is going to help me evolve into the person I want to be?

Meaning provides guidance for answering these and other questions. Of course, this raises questions about what it takes to find this lasting guidance. I am convinced that curiosity is the ultimate tool you need to knit together the future you want.

Of the everyday people I interviewed over the course of writing this book, one person stands out in my mind as having a unique meaning-making capacity. I wasn't the only one captivated by his approach to life; I first learned about Lee Wheeler in a profile that appeared in the *Washington Post* that described him as a sculptor who acquired the highly unusual interest of collecting artwork created by serial killers. In the article, the reporter asked him to share his attitude on life.

I like the idea that if someone is curious enough they'll find what they're looking for. One of those things that I found was not something that I could actually pick up and take with me. I noticed that the lawn had this big bubble in it. It was like somebody put a pillow under the grass. Pushing the mower over it, you notice something like that. What in the hell. I decided to poke my finger down in there. I pulled my finger out, the worst stink in the world . . . blew my hair back. And it was gone. You don't know what to say. You don't know what to do. Your lawn just farted on you. And you don't know whether to be overjoyed, start laughing, or throw up.

Burrowing into places where the average person wouldn't trespass, Lee exemplifies the role of curiosity in finding everyday meaning and creating more of it on demand. His searching wasn't driven by discontent. Rather, by nature he was open and receptive, and added to his meaning supply in the same way many of us routinely supersize our fries when asked. As an artist, he felt ill at ease categorizing himself as a sculptor, "Let's just say that I'm comfortable doing anything from foam rubber to concrete. There are ways to using almost any material. Even if I have to stretch myself, if I like the idea behind the work someone offers to me, I take it on and learn as I go."

Lee yearns to figure out how things work. He gets his hands dirty instead of relying on the so-called wisdom of experts and gurus. His approach is no different than a young child seeing things for the first time without preconceived judgments. With this mindset, he finds ideas for his artwork and he continually adds new terms to define himself. He finds pockets of meaning that help him make sense of himself, other people, other things, and ways in which he is linked to them.

Try as we might, it's not possible to talk about making, finding, and creating meaning without using words like *explore*, *learn*, and *discover*. Think about it, if you don't explore, dig, and search for meaning, how are you going to find it?

Whether you're young and exploring life, feeling unfulfilled, dealing with loss or a major change, if you aren't open to discovery, you won't have access to new experiences and feelings you can capitalize on. This is what we can learn from the Lee Wheelers of the world. Searching and exploring requires attention, awareness, and effort. I can't imagine a better use of our resources than to fill up on meaning to develop the core of our identity.

As my own story attests, we can find meaning and grow even during the difficult times when our beliefs about the world are disrupted by pain and loss. But you don't need major life upheavals and trauma to find meaning. There are an infinite number of places to start searching, so many that it's easy to get lost.

Seeking Purpose, Like Beauty, from Within

From reading a book, tending to a garden, having a metaphysical epiphany, or searching Google for something that was bugging you (When do fish sleep? How can we use volcanoes as an energy source?). From the small details to the big searching questions, we find meaning and make sense of our world and our place in it. Life is meaningful when we "get it."

We regularly experience meaning when we make sense of trivial and mundane events or when we work through more complex and perhaps fraught matters that affect not only ourselves but those we love and care about. Far less often, we experience a level of self-understanding that transcends our routine perspective and transforms us. These insights can have a profound impact on our well-being. This is the sense-making part of meaning. It is about exploring and trying to comprehend who we are and how we fit into our little niche in the universe. It is about understanding the significance of our lives and the other people and things in them.

Purpose is a particular type of meaning. It's when we find a mission or a philosophy of life that we abide by. For instance, Gandhi is an exemplar of purpose in life. He had at least three: to achieve personal enlightenment, help his fellow citizens do the same, and protect the freedom and dignity of human beings (particularly the downtrodden and oppressed). Not too shabby for one man.

Besides Gandhi, plenty of other people devote their lives to protecting fellow human beings who can't protect themselves. Having this type of purpose naturally leads someone to create certain goals and not others. For example, one person might decide to become a police officer to uphold justice and fairness. Another might end up adopting a child from an orphanage. Yet another might work for a living to support their family and donate money to charities to help others in need. People who live purposeful lives have well-defined goals and enjoy devoting considerable effort to accomplish them.

Defining Purpose

There are several hallmarks of a life defined by purpose. **When purpose is a guiding force, there is a central theme in a person's life narrative or story.** If you think of your identity as a building, purpose is the architectural framework that provides an understanding of the totality of who you are. Goals such as completing a master's degree or becoming a better listener with close friends provide an understanding of small spaces in the building—pillars, floors, and rooms. Goals provide an understanding of what we are about at a particular time and place in our life but they miss the big picture.

For instance, if you talk to Carlos, a criminal lawyer, when he's engrossed in a big case, he will describe his plans for keeping his client out of jail. Being a lawyer is an incredibly rewarding career for him. He's proud of his contributions, and it allows him to use many of his strengths. However, when it comes down to it, his purpose is caring for his loved ones. His firm knows not to interrupt his family time unless there is dire need. He is content being with his wife and two kids. They ground him; they bring him equanimity in a chaotic world. The money and benefits of being a lawyer provide security and well-being for his family. His work is important and meaningful, but it is secondary to being a doting husband and father in his life narrative.

Carlos's purpose provides the basis for defining goals and making decisions in everyday life. When choices have to be made, he errs on the side of caring for and being with his family above all else. His purpose provides a sense of direction for investing limited time and energy to engage in, persevere at, and make progress toward particular life goals. When his partner and kids want to go for a weekend getaway, he puts in extra hours at night during the week so he can go with them. It's painful to lose sleep, but he doesn't suffer. As the adage goes, "the juice is worth the squeeze."

Purpose is a manifestation of our core values and interests (see Chapter 5). Instead of governing behavior, purpose offers direction just as a compass offers direction to a navigator. A person may choose not to follow that direction. Having a purpose is just a starting point. Only by committing effort do we give ourselves a self-sustaining source of pleasure and meaning. Without planting the seeds, you don't get the benefits of plants and food (you "reap what you sow").

If Carlos regularly works 90 hours a week and doesn't get home until his family is asleep, he gains little by having a purpose of caring for and sharing his life with them. When we act in ways that run counter to our purpose, we suffer. Gandhi talked about days when he failed to act in an enlightened way. He was human and sometimes he would let his temper get the better of him and scold his wife and kids. When he recognized this gap between his purpose and his actions, he suffered.

When purpose is pursued, a person devotes more effort to important goals and activities, yet it often feels effortless. The same amount of effort is draining when we are doing something that we feel we have to do. Having a purpose protects us because having something important to fall back on that is under our personal control allows us to absorb the strains and difficulties of the world around us. Having a purpose is the firm foundation that allows us to take risks, explore, and find and create meaning. Purpose increases our energy supply to do more of what we want and extract more from it.

When purpose is a guiding force, a person behaves relatively consistently in all contexts, both public and private. We can't legislate purpose. Some parents are going to view taking care of their children as an important source of meaning and pleasure but not their purpose in life. Even if parenting isn't at the core of their identities, this is nothing to be ashamed of as long as they take their role as caregiver seriously.

Just because mothers or fathers view parenting as a purpose doesn't mean there aren't moments when they are exhausted or wish they were free to go away for the weekend to soak in a hot tub and laugh with adults (only adults). They will. However, parents with purpose are more resilient to stress than mothers who view parenting as something they should do or have to do, instead of want to do for its own sake.

Basically, by being attuned to your innermost values and interests, having a purpose makes it easier to create important goals and make inroads toward succeeding at them. When you are doing things that are central to your identity, everything feels less difficult, and stressors take less of a toll on your body, mind, and social life.

Three Paths to Purpose in Life

There are certain venues in which we are most likely to see purpose in action (e.g., religion and spirituality, work as a "calling," being a parent, and caring for other people). Besides work, a common thread is the use of one's strengths in the service of something larger than the self. These aren't the only areas where a sense of purpose can exist. Plenty of people find purpose from hobbies, interests, and passions. There is no single type of purpose, and there is no particular time in your life when you are destined to find it.

I have learned that there is no single approach for creating a purpose. But my colleague, Patrick McKnight, and I believe we can define at least three paths to developing a purpose in life.

1. **We can learn,** modeling our behavior on that of other people and adopting their values and sources of meaning as our own.
2. **We can respond to life-altering events** with a mindset of openness and compassion. By doing so, we can make meaning and add clarity in the aftermath.
3. **We can seek,** intentionally opening doors, searching for purpose.

By presenting these paths, I hope you will realize your potential for creating purpose. With a curious mindset, you will be more likely to stumble upon purpose and put in the necessary effort to solidify a purposeful life.

Learning Purpose from Others

As the first path to finding purpose, we can learn through observing, imitating, and modeling others. We watch others, taking note of the costs and benefits of what they do. If you see your brother grab a great spot underneath a parked car during a game of hide-and-go seek, you may wonder why you didn't think of it. Later, when you hear the engine turn on and watch your brother frantically run for safety, you will never think of it as a hiding spot again.

When you think about the most common areas of purpose in life—work, religion/spirituality/faith, and caring for other people—it is easy to envision how these paths were adopted by following learned rules and guidelines. After all, it is much more efficient to learn by copying than by going on a self-discovery quest fraught with risks, failures, uncertainty, and anxiety. Although it might be more efficient, there is the risk of adopting a purpose too early, long before you discover your own strengths and interests. If we commit to a purpose too early in life when we are vulnerable, easily swayed by charismatic or forceful people in our lives, we could be doing what they want us to do instead of what we are interested in for its own sake. An important mantra to continually ask yourself is who you are committing to a mission for? If I waved a magic wand so that you were guaranteed a lifetime of being accepted, loved, and admired, what would your life be about? What do you stop (now that you no longer worry about what others think)? What are you going to do differently? If your quest would be different, then it's time to begin the journey of your true self. Keep asking these questions over and over again to ensure your purpose is an outgrowth of your own interests.

Children observe their parents and mimic their parents' behavior.

Children go on to eat many of the same foods and develop many of the same habits and preferences as their parents. But parents are just one type of role model.

I listened to a man in his 40s recount a moment in the streets of Manhattan when he watched a stranger stoop down to aid a dehydrated, homeless man lying facedown on the sidewalk. Before this man stopped to help the man, others passed by, ignoring him. Yet as soon as this stranger intervened, the attitude changed and others felt compelled to stop and help out further. He never forgot the incident. It was a crash course on the power of one person to create a better world. When you see that kind of act, it affects what you might do in future crisis situations. The more an observer sees acts similar to that one act and recognizes the benefits of being kind and generous, the more likely a person will add these behaviors to their own repertoire.

For some, purpose originates from others. That is to say that the purpose did not come from a laborious act of self-discovery or the chance occurrence of a transformative life event, but rather from the mere observation of another's behaviors and the associated emotional reaction that was paired with those behaviors.

It begins by observing others. Through observation, people are able to mimic the behaviors for others to see. Spreading through other people like a viral transmission, purpose can take hold of an open-minded or vulnerable body. Religion serves as an excellent example of social learning. With respect to religion, children who grow up in observant households watch their parents, siblings, friends, and neighbors behaving according to religious faith. Parents show their kids what to value and teach them to believe certain things and not others. Those beliefs and behaviors become "normalized" and soon are acted on by the kids. The more a community fosters these behaviors, the stronger the religious following becomes. In a sense, religion spreads through others.

In the 2007 Pew Religious Landscape Survey of 35,556 Americans, less than 10% of American adults reported being without a religion as a child. Over 90% reported a childhood religion. Clearly, few parents allow their children to remain unaffiliated until they are older and in a stronger position to decide what beliefs they want to adopt at the core of their identities.

For some children, religion no longer becomes the province of their parents' world. It becomes the center of their own, guiding them as the central structural framework of their life. That is, affiliating with a religion early in life causes a purpose to develop. Once it becomes a purpose, it takes on a life of its own, guiding decisions on what to do when faced with multiple choices. For instance, do I put my overflowing income into the bank or do I follow the biblical tradition of a tithe and donate 10% of my money to the church or a charity? If you belong to a church and feel that it is your mission in life to follow your religion and God, then your actions should match your beliefs.

The key question is whether you live in accordance with your purpose or your purpose is superficial window dressing. Think about this the next time you describe the core ingredients in your own life, whether it is religious or spiritual beliefs, parenting, or whatever mission you believe is the centerpiece of your life. Do your actions match your beliefs? How much effort do you devote to pursuing goals related to these beliefs? If there is a gap between what you say and do, figure out why.

We don't get to pick our parents, and we have little say about much of what we are exposed to early in life. If you are lucky, you get a good batch of caring, protective parents who provide you with substantial autonomy to make choices and discover your personal strengths and interests. If you are less fortunate, your life gets carved out for you. Maybe your parents were abusive, maybe something bad happened to you, maybe you never learned about the wide variety of options at your disposal and that you can change directions anytime you want.

Modeling other people doesn't end with our parents. We hear of purposes by talking to people or being exposed to ideas via television, movies, books, classes, the Internet, and countless other information sources. When you look beyond your parents for guidance, be sure to identify people who possess the success you want. They may have strengths that you possess except theirs are fully developed. They may have strengths that you don't possess that impress you. You don't have to choose the chance offerings in front of you such as parents and teachers. Be curious, open, and proactive in your search for good role models and learn from them while staying attuned to your uniqueness. When you discover your own strengths, when you are at your best and the most energized, you will be similar and different from even the best-fitting model.

There are few rules and blueprints in our freedom to find the purpose that will guide our own unique lives. I bring this up because when it comes to learning from others, it's easy to be mesmerized and obedient. When we put our curiosity on hold, we risk losing ourselves.

Reacting to Life When Opportunities and Challenges Intervene

The first noble truth of Buddhism is that we are going to encounter hardships. Besides feeling pain, these events often alter our very assumptions about ourselves, other people, and the world around us. Yet, in spite of the often overwhelming obstacles people face, a large number of children, men, and women are able to absorb the pain and continue moving forward. That is, they are resilient. Others, in dealing with hardships, actually bounce forward. That is, they grow and benefit from their struggles.

To give you an idea of the science behind resiliency and growth following trauma, let's consider one of the most horrifying things that one person can do to another: rape. As you might expect, rape victims are at a dramatically greater risk for emotional disturbances than people who have never been a crime victims. For instance, rape victims are 6 times more likely to develop post-traumatic stress disorder at some time in their lives and 13 times more likely to attempt suicide. Responding with anxiety, depression, nightmares, and thoughts about suicide are normal reactions.

But one truly amazing, less-well-understood fact is that after six months, without therapy, about 50% of rape survivors bounce back to how they were feeling beforehand. Many survivors don't develop mental or physical problems, and they show only minor impairments in their ability to work, socialize, and take part in their regular routines.

We see the same thing when combat veterans return home. Some witnessed horrific violence and atrocities and some were themselves wounded and almost killed. On the news, you hear about an epidemic of post-traumatic stress disorder, depression, suicide, and an inability to function as contributing members of society. These after-effects are devastating and all too real and urgent for the people affected directly and society as a whole. But without minimizing the impairment of a significant minority of veterans (and believing we should be doing everything possible as a society to express our gratitude for their sacrifice by aiding them), we can learn a lot from the veterans who show a profound ability to cope and thrive despite the hardships of their experience.

This phenomenon, which has been studied in extremely violent contexts, also happens following more common events. A woman told me, "Like it or not, your life gets altered after you give birth." The same goes for when we lose a job, change jobs, move to a new city, retire, fall in love, or get married. Events such as heart attacks, serious car accidents, illness and death of children and loved ones, being diagnosed with severe and terminal illnesses, and watching the events of September 11, 2001 unfold on television are some of the events studied by researchers to understand how people cope and thrive.

From a review of this research, it appears that there are three fundamental ways by which people might grow from transitions, difficulties, and crises:

Relationships are strengthened. There is greater appreciation of the importance of family and friends. Loved ones who didn't receive proper attention before are now elevated to top priority. There might be an increased devotion to caring for other people. Interpersonal virtues such as kindness, generosity, forgiveness, and gratitude emerge. A renewed sense of connection to others and a desire to act upon these feelings is common. We might call this compassion borne out of suffering.

Our view of ourselves changes. Recognition of the personal strengths that helped you effectively cope provides a new source of confidence to deal with future hardships. Wisdom emerges from the recognition of what strengths are possessed as well as a greater acceptance of existing vulnerabilities and limitations. Instead of threats, survivors often start to view difficulties as challenges and even opportunities. This new attitude often leads to greater resilience so that in the future, less distress and upheaval occur in the first place.

Life philosophies are changed. A purpose or mission in life might emerge. Stronger spiritual or religious beliefs often arise following hardships. It could be a renewed commitment to existing beliefs and practices or switching your affiliation to a new group. From this, a stronger relationship forms with a higher power such as God or nature. But a new purpose can have other themes as well. It could be the pursuit of knowledge, making sure that your strengths and creativity are shared with the world, caring for particular people in your life, working toward fairness and social justice in the world, protecting the environment, expressing yourself in the most authentic way possible, and so on. From these life philosophies, more concrete goals are created and worked toward on a regular basis. It is only when these philosophies are transformed into action that we can say that a person is living a purposeful life in the aftermath of stress or trauma. When we move from beliefs to action, we seem to get the most psychological, physical, and social benefits.

What these life-altering events bring into the equation is randomness and chance. A close look at the research and clinical literature gives us an idea of how to improve the likelihood of becoming a resilient warrior who bounces back or someone who grows from these events and bounces forward. Keep in mind that it's impossible to address everything in a single chapter and our focus is on how random, difficult events can set the stage for purpose in life. Based on what we know are the most effective strategies for improving the lives of survivors, let me give you some insights to increase the likelihood of attaining positive (or less negative) outcomes.

- Shift the way you think about trauma and adversity. The research is clear that not everyone ends up with a depleted, damaged existence after experiencing trauma. Many people are resilient, and many people actually benefit from the experience and grow, becoming stronger and more purposeful. Keep in mind that suffering and growth coexist. That is, you might experience profound distress and loss at the same time that you personally transform for the better. In fact, you need negative emotions, thoughts, and distress to grow.

- There is nothing inherently positive about traumatic events and adverse experiences. Any growth that does take place is not because of the actual event. If you find purpose, it will arise from struggling with the new reality in the aftermath. We have to work with the pain, negative feelings, negative thoughts, and unwanted sensations that we are carrying with us in the present. We can't change the past. By taking action and living in the present, we can create a workspace for us to grow.

- Trying to make sense of shattered assumptions and become a fully functioning person requires us to accept that a difficult event happened and difficult thoughts and feelings are normal responses. It is only from this acceptance of the negative that we can try to make sense of what happened and detect meaning and purpose.

- Recognize where you are starting and what you want to do. Everyone's wounds are not equal. From the onset, judging yourself against other survivors for who is doing better is a losing proposition.

- A significant number of people will grow very little or not at all from their hardships. Joseph Campbell expressed it best at the beginning of this chapter: Do not aim to find purpose, aim to feel alive in your moments. If you find purpose, great. Maybe you won't but still, hopefully, you will recover and fill your existence with rich, meaningful moments.

- The people least likely to grow are those experiencing minimal distress and those experiencing tremendous distress. People who experience minimal distress are resilient. They don't need to search for meaning and try to make sense of their experience because they effectively mastered the situation on their own. If it's not broken, nothing needs to change. People who experience tremendous distress are at a different stage. They are trying to survive. Growing is the least of their concerns. They are looking for a sense of safety, they are looking for hope, they are figuring out how to function in everyday tasks that we take for granted: sleeping without nightmares; unexpected intrusive reminders of the event; feeling as if they are reliving the event, and thus, being stuck in the past and unable to participate in the present or work toward a meaningful future.

- If you do not grow from adversity, you did not do anything wrong. Many factors affect whether we grow and many of them are out of our control. The number one priority is safety and functioning in everyday life. Only then can we start thinking about learning and growing from the experience.

- It is normal to experience positive and negative life changes following trauma and adversity. Once again, it is normal for the good and the bad to coexist. I had a client who lost his wife seven years before I met him. He said, "I always sleep on my side of the bed." The loss was physically palpable to him. He found a new appreciation for the little things in life that he never paid attention to before. He created a garden after noticing how often he felt at peace looking at his neighbor's flower beds. At the same time, all of the activities that his wife asked him to join her in are no longer part of his routine. He accepts these gains and losses that add up to his current, changed life space. Interestingly, people who reported both a high frequency of positive and negative changes following the September 11, 2001 terrorist attacks experienced the most growth. Remember that working with the negative will aid you in finding meaning and purpose.

- Besides the intensity of initial distress and intrusive thoughts, two other broad factors contribute to the amount of growth we experience. First, the length of time that has passed since the event took place matters. We need time to comprehend and make sense of what happened and determine whether we are ready to commit to any potential positive changes (relationships, self-views, and purpose in life). Second, curiosity and openness to these experiences is important. Without a motivation to see what changes took place and be willing to move in that direction, any possible growth will cease. It is about seeing things as they are without illusions and blinders, even when it hurts to do so.

- If we sit back and let things happen to us without doing anything, we are less likely to find purpose. This doesn't mean we should be spending the bulk of our time ruminating and questioning what happened. We should be observing, exploring, and discovering. Purpose has an opportunity to arise when we first accept that an upheaval occurred and is part of our history.

- Take action with the unwanted thoughts and feelings in tow. When we recognize a purpose, we reevaluate our priorities about what concerns us, what we are committed to, and how we want to devote our time and energy. To transform our new relationships, view of self, or life philosophy into action, we need to create concrete goals and work toward them. By figuring out how we can put effort toward these goals in a given day, we stitch meaning into a slightly revised identity. The appendix of this book includes some questionnaires you can use to begin the initial introspective work toward purposeful living.

Revising and making sense of difficult life experiences is a process. It is possible for some people some of the time. Reacting and managing uncertainty with openness and curiosity is one of the ingredients at the heart of resilience and recovery. When we are open and curious, we have a greater capacity to grow and thrive in the midst of loss and trauma. But the research is clear that curiosity is just one of several factors that are important in purpose development following trauma and adversity.

You saw this happen on a massive scale after September 11, 2001. After going through a range of emotions, including rage and hatred, many people made radical changes in the way they do business. Newspaper reports were filled with stories of successful investment bankers, lawyers, and plastic surgeons trading their high-status, fine-dining lifestyles to make a difference in the world. There was a sudden influx of recruits wanting to become elementary school teachers or nurses and joining causes such as helping refugees and working as doctors in third-world countries (for a fraction of what they could earn in practice in their home countries). Scientists found evidence of these changes at a psychological level. You could see that on average, Americans became more grateful, kind, spiritual, and interested in new experiences as a result of September 11, 2001.

These studies were done right after the terrorist attacks and unfortunately, we now know that for far too many, those changes didn't last. They returned to their old lives in a few months. Their growth was transitory. But for a minority, their lives remain altered. The small American flag suction-cupped to the window of their cars might be gone, but they modified fundamental life philosophies and values. From this, their goals and everyday actions changed for the better. From this difficult event, they found a new life devoted to things bigger than themselves. They found purpose.

Being Proactive: Lust for the New, Explore, Experiment, and Discover

Dr. Rache Simmons, a breast cancer surgeon at Weill Cornell Medical College, found a purpose by reflecting on her interests and unveiling a career directly linked to them. She describes an apartment "full of naked women. Paintings, sculptures, bronzes." She adds, "As an undergrad, I was an art history major." Ever since she was young, she had "a deep appreciation for the female form . . . that passion has had an effect on me. I do, in a way, look at my patients' bodies as works of art. I do a lot of lecturing, and the main theme today is less and less invasive therapies, which I have been fortunate to help pioneer. Less invasive means less disfiguring. I love it when a lumpectomy patient a year later can't find her scars. Or when a mastectomy patient goes for a mammogram and the radiology tech starts to x-ray the reconstructed breast because it looks so natural." Dr. Simmons does her best to preserve beauty and minimize the pain and suffering in the world by a few degrees. Many people have an aesthetic appreciation for the naked human body but Dr. Simmons took it a step further, unleashing a purpose that honors this interest.

A purpose is difficult to form, and we put our life on hold if we wait for chance events to spur us into action. Like Dr. Simmons, purpose can arrive from a deliberate searching and refining process that I call being proactive. Instead of reacting to transformative events, it is an intentional and gradual process. Curiosity is a critical ingredient.

Enjoy the pleasures of thinking. Healthy introspection allows us to better understand what is at the core of our life narrative. Know what you value, what strengths are in your possession, and how to wield them. By being more aware of what our capacities are, what is important to us, and what interests us, we allocate more of our limited time, effort, and energy in a given day to things that are worthwhile. It is easy to narrowly think of curiosity as exploring the world outside us. Don't ignore the value of also being curious about who you are and the benefits of introspection. We can extract more meaning from the things we do.

Without exploring outside the confines of our life space, our meaning is limited. Curiosity attracts people to new experiences. These ingredients guarantee exposure to novelty. Self-expansion is inevitable when a person is exploring the unknown or challenging the limits of their knowledge and skills. This expansion process can include clarifying and strengthening preexisting interests and values (depth), or increasing the number of interests and values that we use to define us (breadth). By immersing ourselves in new expansive activities, sometimes an interest, hobby, or passion develops, which can be defined as reliable sources of joy and meaning that are intentionally invested in. These sources of enjoyment and meaning are important, but they often possess little connective tissue with the other elements in our personalities. That is, they are not necessarily a purpose in life.

To fully develop a purpose, we must be able to recognize and capitalize on situations that allow for illuminating experiences (those aha!

moments). We have to take time out of our busy lives to reflect on, make sense of, and integrate the new. We can't always be seeking novelty. We also have to make sense of it and see if what we experienced is a momentary adrenaline rush, a one-night stand of meaning-making, or a new addition to our identity. If it is a new addition, we need to figure out whether it is another brick in the wall or something much more central to the entire architectural framework—what we call lasting meaning or purpose.

Introspection, self-knowledge, and curiosity are not the only requirements for being proactive to develop purpose. A final requirement is chance or serendipity. Chance often plays a role in recognizing purpose. There exists some randomness in what we get exposed to in our everyday life. Think of the perfect way to meet a spouse—two strangers grab for a cantaloupe at the supermarket at the same time, they turn to face each other and giggle over a shared moment, and the next thing you know they are trading phone numbers and a few months later, living together happily ever after. If we regularly explore, experiment, and play, sometimes we are going to stumble upon positive chance encounters. The more you try, the more chances you have to find something meaningful.

When that serendipitous or chance event occurs, we can make better sense of it and strengthen it by talking to other people. We make capitalization attempts—sharing our good news and, hopefully, getting constructive responses. This only intensifies our initial interest and excitement. We are more likely to invest in a chance encounter when someone we trust and rely on validates our ideas. But of course, some of us are confident doing this alone and just need ample introspection time. What is most important is that you are receptive to chance encounters and are willing to tolerate the uncertainty and dive in to see what rewards are there for the taking. Chance might not be under our control, but we can regulate our attention and the quality of our attention to be mindful. Through trial-and-error, intentional reflection, and curious exploration, we can build strong and broad structural additions to the self from random positive moments in our daily lives.

At some point in the trial, error, and chance process, we may come to the realization that a satisfactory purpose has been reached (or at least a purpose that can be further refined). In the process of developing a purpose, the "satisficing," or good-enough, solution is less about pursuing happiness and more about discovering a way to finally express those elusive innermost values and interests.

Going back to that 2007 Pew Religious Landscape Survey, we find a sizeable number of people who, in their search for meaning, found that they outgrew their beliefs. Of adults who were not raised to be religious as a child, 4% later identified with a religion. That is, they found a good fit for what they believe in and what they want their lives to be about. From what we know about people who are religious, we can be sure that some of them stumbled upon a purpose in life. Of course, it goes both ways. Of adults who were raised religious, nearly one in ten decided that this was not the guiding force in their lives and changed their life philosophies to something other than religious faith. Before you write off these changes as too small to be noticeable, consider that in the latest U.S. census, more than 85% of the country reported being religious—10% of this group is more than 25 million people!

A dynamic process is at work, and people are on a quest for meaning. At any time, we can "learn" another set of values to guide us, and from these experiences, we can reshape our brains and modify our personalities.

The Intentional, Gradual Process of Purpose in Action

Going back to Lee Wheeler, the sculptor I read about in the *Washington Post*, I was so intrigued by his farting lawn and curious attitude that I had to talk to him and find out more. It became clear that he exemplifies the proactive mindset.

Lee Wheeler happened to find his own, odd purpose to appreciate and try to understand the absurdity of life and complexity of human beings. Now, it's easy to see how this sort of purpose can lead one to become a psychologist, an international journalist, or any number of professions. Lee's purpose is not limited to his career as a sculptor and artist; it permeates his way of life.

Lee has gray hairs sprouting from his beard in every direction. His eyes gaze intently from beneath a well-worn baseball cap. He speaks of his love of "things being in places they shouldn't be." This philosophy of life is his purpose, his compass.

It's not something you are looking for. It could be a squashed bottle cap that formed a little star. It could be a paper cup that when you look inside the cup it says "fuck you" . . . these are the things that make me smile. You stop and investigate, and it stops the rest of anything else I might have been concerned about. I wasn't focused on taxes or whether my face was breaking out . . . in a lot of my personal art, there is a lot of hidden stuff. When I was in college, I used to write little messages like under the sink, inside of the closet . . . I wanted someone to think "now, what the hell was he doing in here." To be honest, I still do it. It's like a little time bomb. You just leave them everywhere.

What interested me the most was his interest in serial-killer art. I didn't know what this was and from his animated voice, it became clear that I asked about his passion.

I collect interesting things. Maybe not normal things. There are things you pass by and you don't know what its use is . . . I collect the work of serial killers. They aren't known for their art. It's their compulsion to draw—a need to express themselves while they are contained on death row, for example. In a certain way, it's almost like having a conversation with someone. When I land their work, I get a piece of their brain. Sometimes it creeps me out—feeling the horror of what they did. But it still intrigues me. Enough so that I am drawn to their work. One of my favorite pieces was done by Henry Lee Lucas. He abducted just about anybody. It's his lack of emotion and seeing people as subjects and not living, breathing human beings that interests me.

He drew a cardinal and it happens to be pretty well done. No indication of his other side is present. It's a sweet picture you might expect from an old man working in a Woodstock attic strewn with flowers. It's out of context, and I like knowing both extremes.

If you are one of the millions who read mystery novels or watch television crime dramas such as *C.S.I.*, you might find a kinship to Lee. He spends his days trying to fathom how certain people blur the boundary between man and beast to abuse and kill others. Lee is puzzled by people who inflict pain without remorse. His collection of oddities reflects this mission to understand the truth, however complex it is, about the world around him.

He is on an endless expedition to find things that don't fit into boxes and categories whether it's his lawn bubbling with methane gas or serial killers expressing some deep sensitivity within them. Lee experiences moments of intense flow and excitement even if there are lulls in between. When talking, he sounds like a devious child at play. I say this because as adults, we forget about the value of having fun and playing. I suspect this is why he finds pleasure and meaning in places that most of us would avoid or obliviously pass by. Lee is on a trial-and-error quest, and he loves the journey. Receptive to chance encounters, he grabs pockets of meaning and adds them to his life space.

Purposeful Living and Self-Doubt

There is an important barrier to being proactive to find purpose and that's self-doubt. It doesn't matter if you are thinking about changing your career, ending an important relationship, or training to become a triathlete, inside your head is mental chatter about how your goal is ridiculous, selfish, risky, and possibly dangerous. Worry and the need for control are often lurking in the background.

It's hard to turn the chatter off, and when we try to, we only think more about why we should stick with the status quo. It's wrong to assume that if we think our newfound ideas are stupid, we should listen to "intuition." It's wrong to think that we would be more confident and without any doubts if our ideas had merit. I want to emphasize that self-doubt, the need for control, and the fear of uncertainty are all normal, healthy parts of taking risks.

When we are willing to tolerate the notion that what we think we know may be wrong or half-baked, when we are willing to remain open to what we don't know, we become liberated. We become receptive to being on the prowl for lasting meaning and purpose. That tolerance of uncertainty is about remaining open to what your mind has to say while continuing to move in the direction of core values and interests. We all grow up learning about what is appropriate for someone of our age, gender, social status, appearance, and so on. These standards of what we "should" or "ought" to do only get in the way, closing us off prematurely from opportunities that are a good fit for us (in some ways similar to all other people, in some ways different from all other people, and in some ways unique from any other person).

If you close yourself off too early, you are less likely to find purpose. Remain proactive and wield your curiosity.

A Life of Wonder

No matter what path you end up on—and it could be an odd combination of learning from others, being reactive, and being proactive—always questioning, investigating, and wondering will serve you well. I am reminded of the story of a father who tried to drill this mentality into his son, who was still just a naive, inquisitive sponge.

Richard's dad played a game with his son called "Where am I?" Richard closed his eyes and his dad would ask him to guess his whereabouts: "Whatever direction I turn, all I can see are these huge fuzzy blue towers. They're enormous. They are so tall and close together that I can't tell whether they rise straight into the clouds. No matter where I walk, all I see are the same fuzzy blue towers over and over again. Every tower is exactly the same distance from the next one, and every tower is the same width. I just rubbed against one of the towers, and it scratches my skin. No flesh wounds, it feels like the backside of a sponge. When I look at my feet, I notice that the ground is exactly the same as the towers. The ground is fuzzy and blue just like the towers. If I jump up and down, my feet bounce a bit. Where am I?"

The answer was that Richard's dad was in the blue carpet in the bedroom they were sitting in. They placed this game regularly. His goal was to get his son out of the habit of seeing the world from a single perspective. The world was different depending on your starting point and he wanted Richard to learn that there were always different angles and perspectives when looking at a person, place, or problem. If you stuck to one perspective, you were blinded from learning anything new. This game was just one way that Richard's dad worked to develop his son's curiosity, flexibility, and creativity.

Years later, Richard thinks about how his dad influenced him. He credits him for turning him on to science. Richard was quite successful, graduating from a top college, getting a doctorate in physics, serving as the youngest member of the Manhattan Project (the team that built the first atomic bomb), and eventually winning the Nobel Prize. Richard Feynman's life is a testimonial to the value of being curious. His father taught him this at an early age, and it became the fundamental, core ingredient of his life. He describes the influence of his parents that he didn't fully appreciate at the time in his aptly titled autobiography, *The Pleasure of Finding Things Out*.

In dozens of countries, curiosity is one of the most commonly endorsed strengths. Unfortunately, stating that you are curious is not enough to access the profound benefits discussed in this book. It's easy to take your strengths for granted. It's easy to forget about strengths that appear regularly. You don't need to act on or actively attend to them because they just seem to be there.

You might have responded in ways that many of my students do at the outset: "I don't need to try to be a curious explorer because if something is interesting, then I'll react to it." It's easy to underestimate strengths when they are tightly woven into our identities. Maybe we think we should add more strengths to our toolbox than work with the ones we already have. Let me be candid and tell you that this is flat-out wrong. Our strengths are exactly what we should be playing with. These are the behaviors we are most comfortable with; these are the behaviors that are the most energizing; these are the behaviors that feel effortless to use.

When we take curiosity for granted, we don't use it in the service of anything, much less "something larger than the self." Instead of using our curiosity to guide us toward what is intriguing and valuable, we get trapped in well-worn habits and routines. We become passive and wait for new things to happen to us. When we are passive, we limit the amount of meaning that we find and create in our lives. For curiosity to contribute to a meaningful life, we need to intentionally wield it. We need to start paying attention to how we are connected to the world around us (see Haruki Murakami's quote at the start of this chapter). We need to consider the value of being proactive even when we are reacting to random, chance life events.

Being a curious explorer can open the gates to making, finding, and creating meaning. It only happens when we intentionally listen to and act on our curiosity. When we do, we get a taste of what our lives can be when we let our desire to explore reign free.

For some of us, passionate pursuits become the center of our identity and our purpose or "calling" in life. Our actions are aligned with our deepest, most central values. Having a purpose is one of the paths to a healthy, satisfying, and long-lasting life. It is not the only path, but it is an important one. My hope is that by understanding the science behind purpose, some of you will find meaning, some of you will find purpose, and everyone will construct a life that is more open, more exciting, more expansive, and longer lasting than before.

Appendix
Exercises and Tools

Relationship Exercises

In Chapter 6, I discussed the scientifically based idea that embracing and bringing novelty and uncertainty into a relationship can lead to greater satisfaction for both partners. The following four-step exercise is one way to improve the excitement quotient and, therefore, the satisfaction in your relationship.

Exercise 1:
How to Invigorate Your Relationship with Your Romantic Partner

STEP 1: Privately, each person should think about time spent with their partner. Without talking about it, each of you should make a list of the shared times together that could best be described as "very pleasant" or "exciting." Think about things you do at home, for work, in the community, for leisure, on vacation, or anywhere else where you did something with your partner that made you feel excited. For instance, think about when the two of you:

- Went to a concert or a club
- Played or watched a sport or games of some kind
- Shopped
- Learned a new skill
- Talked
- Volunteered
- Solved a problem

- Took care of other people, animals, or things
- Went to a spiritual or religious event/workshop/meeting
- Played music
- Had sex (the more details, the better)
- Worked out
- Relaxed
- Spent time in a different environment than you are usually in (beach versus mountains, suburbs versus city, noisy versus quiet, teeming with people versus sparsely populated)
- Engaged in strenuous physical and/or mental exercise
- Joined an organization that you both believed in
- Pursued a hobby
- Worked on the house, the yard, the car, the boat
- Cooked new recipes
- Went to the movies
- Sat in the same room and did your own thing, like read, did needlework, or worked crossword puzzles
- Planned the family budget
- Took a class
- Something else (the sky is the limit—add any activities that fueled you)

STEP 2: Get together with your partner and compare the two lists. Take note of which items show up on *both* lists. Create a master list containing those items.

STEP 3: This step has three parts. First, think about some activities that are similar to the activities you found exciting and are possible for you to do, and add them to the master list. Second, be on the lookout for new, novel activities that you believe both you and your partner will find exciting, and try them out. If your instincts were right, add them to the master list. Third, consider incorporating activities that intrigue you that you might ordinarily pass up because you feel they are too childish or have a potential to be embarrassing. Stretch by allowing yourself to feel some anxiety. As you discover new, exciting activities, add them to the master list.

STEP 4: Engage in one of the activities on your master list for at least 1½ hours every week. Add spice to these occasions by being sure to select examples from your master list of the three different types of activities discussed in step 3. After all, even the most exciting thing will start to feel mundane if you do it over and over again, and sometimes it takes a few tries before you and your partner find activities that are the right fit for you. Treat this 1½ hour period each week as a very important appointment—don't skip it unless you absolutely must.

When You Are Ready for the Next Level

You probably remember from Chapter 6 that doing something novel and exciting with your partner *and* close friends is another great way to enhance a relationship. The research shows that people you care about often influence and inspire you in positive ways. Additionally, it can be interesting to see parts of your partner's personality emerge in ways that you don't often see when it is just the two of you. Here are three additional steps for those of you who want to maximize the potential for developing a partner relationship that has energy and excitement.

Exercise 2: Three More Steps

STEP 5: After at least two months of experimenting with new activities together, create a second master list with your partner. This list should have two columns. In the first column, write the names of friends with whom you both like to spend time. In the second column, write the things that you do with these friends that you find "very pleasant" or "exciting."

STEP 6: Tell your friends that they are important to you and you would like to see them more often. Brainstorm with them about what kinds of activities would be most enjoyable.

STEP 7: While continuing to engage in activities that are novel and exciting with just your partner every week, plan some outings with friends. Go out, and have fun!

Measurement Tools

In many of the studies that provided the scientific foundation for this book, psychologists often ask people to answer questionnaires about certain topics. The answers that people give, when they are being honest, often provide tremendous insight into what that person thinks, feels, and does.

The measurement tools that psychologists use are not something that someone just made up and printed. Instead, they have been given to hundreds, or even thousands, of people in a process called "validation." Validation allows the researcher who developed the tool to know that it measures what it is supposed to measure and not something else. The process of validation occurs before the tool is used outside of a research laboratory.

To help you understand yourself better, I received permission from several well-known and highly regarded researchers to reprint their measurement tools, as well as reprinting one of my own. So work your way through them, and enjoy the process of getting to know yourself better. It is important to highlight that there are no right and wrong answers, and you are not competing with anyone. The more honest you are in responding, the more you gain in understanding yourself.

The Curiosity and Exploration Inventory-II

The Curiosity and Exploration Inventory-II uncovers the degree to which you are a "curious explorer." Read each statement and circle the number that best matches the description of how much you agree with the statement. Do not think about how you wish you could be, or how you might be in the future, but how you actually are right now. Be honest with yourself.

1	2	3	4	5	
very slightly or not at all	a little	moderately	quite a bit	extremely	

		1	2	3	4	5
1.	I actively seek as much information as I can in new situations	1	2	3	4	5
2.	I am the type of person who really enjoys the uncertainty of everyday life.	1	2	3	4	5
3.	I am at my best when doing something that is complex or challenging.	1	2	3	4	5
4.	Everywhere I go, I am out looking for new things or experiences.	1	2	3	4	5
5.	I view challenging situations as an opportunity to grow and learn.	1	2	3	4	5
6.	I like to do things that are a little frightening.	1	2	3	4	5
7.	I am always looking for experiences that challenge how I think about myself and the world.	1	2	3	4	5
8.	I prefer jobs that are excitingly unpredictable.	1	2	3	4	5
9.	I frequently seek out opportunities to challenge myself and grow as a person.	1	2	3	4	5
10.	I am the kind of person who embraces unfamiliar people, events, and places.	1	2	3	4	5

Statements 1, 3, 5, 7, and 9 describe stretching, or how much a person challenges himself or herself to learn new things or experience new things. The higher the score, the more you look for and enjoy challenge. Out of a possible 25 points, my score is ____.

Statements 2, 4, 6, 8, and 10 describe embracing, or the degree to which a person enjoys the uncertainty associated with participating in new activities. The higher the score, the more you embrace uncertainty and mystery. Out of a possible 25 points, my score is ____.

CEI-II: Kashdan, T. B., Gallagher, M., Silvia, P., Breen, W. E., Terhar, D., & Steger, M. F. (2008). The Curiosity and Exploration Inventory-II: Development, factor structure, and initial psychometrics. *Manuscript submitted*.

This is an improvement over the initial, well-validated version: Kashdan, T. B., Rose, P., & Fincham, F. D. (2004). Curiosity and exploration: Facilitating positive subjective experiences and personal growth opportunities. *Journal of Personality Assessment, 82,* 291–305.

Basic Psychological Needs Scale

The Basic Psychological Needs Scale measures several elements of well-being, that is, the degree to which a person believes their needs for autonomy, competence, and sense of belonging are satisfied.

Autonomy refers to feeling that one's choices and activities are self-determined as opposed to being controlled by internal (such as guilt) or external pressures (such as other people's rules).

Competence refers to feeling a sense of confidence and mastery in one's environment.

Belonging refers to feeling that satisfying and meaningful connections are being made with others and you feel cared for, validated, and understood by others.

To find how well you satisfy the three basic needs, read each statement and circle the number that corresponds to how true the statement is to you.

1	2	3	4	5	6	7
not true at all			somewhat true			very true

		1	2	3	4	5	6	7
1.	I feel like I am free to decide for myself how to live my life.	1	2	3	4	5	6	7
2.	I really like the people I interact with.	1	2	3	4	5	6	7
***3.**	Often, I do not feel very competent.	1	2	3	4	5	6	7
***4.**	I feel pressured in my life.	1	2	3	4	5	6	7
***5.**	People I know tell me I am good at what I do.	1	2	3	4	5	6	7
6.	I get along with people I come into contact with.	1	2	3	4	5	6	7
***7.**	I pretty much keep to myself and don't have a lot of social contacts.	1	2	3	4	5	6	7
8.	I generally feel free to express my ideas and opinions.	1	2	3	4	5	6	7
9.	I consider the people I regularly interact with to be my friends.	1	2	3	4	5	6	7
10.	I have been able to learn interesting new skills recently.	1	2	3	4	5	6	7
***11.**	In my daily life, I frequently have to do what I am told.	1	2	3	4	5	6	7
12.	People in my life care about me.	1	2	3	4	5	6	7
13.	Most days I feel a sense of accomplishment from what I do.	1	2	3	4	5	6	7
14.	People I interact with on a daily basis tend to take my feelings into consideration.	1	2	3	4	5	6	7
***15.**	In my life I do not get much of a chance to show how capable I am.	1	2	3	4	5	6	7

continued

*16.	There are not many people that I am close to.	1	2	3	4	5	6	7
17.	I feel like I can pretty much be myself in my daily situations.	1	2	3	4	5	6	7
*18.	The people I interact with regularly do not seem to like me much.	1	2	3	4	5	6	7
*19.	I often do not feel very capable.	1	2	3	4	5	6	7
*20.	There is not much opportunity for me to decide for myself how to do things in my daily life.	1	2	3	4	5	6	7
21.	People are generally pretty friendly toward me.	1	2	3	4	5	6	7

* Statements that are worded in a negative way must be reverse-scored, so figuring out your score for this tool requires several steps:

STEP 1: Reverse-score statements 3, 4, 5, 7, 11, 15, 16, 18, 19, and 20. That means if you circled a 1 for any of those statements, change it to a 7. Change a 2 to a 6, a 3 to a 5, a 5 to a 3, a 6 to a 2, and a 7 to a 1. If you circled a 4, keep it as a 4.

STEP 2: Add the scores for statements 1, 4, 8, 11, 14, 17, and 20, and then divide the total by 7. That is your average "Autonomy" score. My Autonomy score is ___.

STEP 3: Add the scores for statements 3, 5, 10, 13, 15, and 19, and then divide by 6. That is your average "Competence" score. My Competence score is ___.

STEP 4: Add the scores for statements 2, 6, 7, 9, 12, 16, 18, and 21, and then divide by 8. That is your average "Relatedness" score. My Relatedness score is ___.

The Basic Psychological Needs Scale: Deci, E. L., & Ryan, R. M. (2000). Self-determination theory website, Basic Psychological Needs Scale-General. http://www.psych.rochester.edu/SDT/measures/needs.html.

Validation for the Basic Psychological Needs Scale is provided in the following published articles:

Gagné, M. (2003). The role of autonomy support and autonomy orientation in prosocial behavior engagement. *Motivation and Emotion, 27,* 199–223.

Kashdan, T. B., Mishra, A., Breen, W. E., & Froh, J. J. (in press). Gender differences in gratitude: Examining appraisals, narratives, the willingness to express emotions, and changes in psychological needs. *Journal of Personality.*

Niemiec, C. P., Ryan, R. M., & Deci, E. L. (in press). The path taken: Consequences of attaining intrinsic and extrinsic aspirations in post-college life. *Journal of Research in Personality.*

In these articles, you can find out how your score compares with the typical scores of other people from around the world.

Acceptance and Action Questionnaire-II (AAQ-2)

The AAQ–2 measures psychological flexibility or the process of being alive in the present moment, working with any unpleasant thoughts, feelings, and bodily sensations (instead of avoiding them), and moving in the direction of important values. To see where you score, circle the number next to each statement that best corresponds to how true you believe the statement is for you.

1	2	3	4	5	6	7
never true	very seldom true	seldom true	some- times true	frequently true	almost always true	always true

1.	Its okay if I remember something unpleasant.	1	2	3	4	5	6	7
*2.	My painful experiences and memories make it difficult for me to live a life that I would value.	1	2	3	4	5	6	7
*3.	I'm afraid of my feelings.	1	2	3	4	5	6	7
4.	I worry about not being able to control my worries and feelings.	1	2	3	4	5	6	7
*5.	My painful memories prevent me from having a fulfilling life.	1	2	3	4	5	6	7
6.	I am in control of my life.	1	2	3	4	5	6	7
*7.	Emotions cause problems in my life.	1	2	3	4	5	6	7
*8.	It seems like most people are handling their lives better than I am.	1	2	3	4	5	6	7
*9.	Worries get in the way of my success.	1	2	3	4	5	6	7
10.	My thoughts and feelings do not get in the way of how I want to live my life.	1	2	3	4	5	6	7

* Statements 2, 3, 4, 5, 7, 8, and 9 must be reverse-scored. Looking back at those six statements, if you circled a 1, change it to a 7. Change a 2 to a 6, a 3 to a 5, a 5 to a 3, a 6 to a 2, and a 7 to a 1. If you circled a 4, keep it as a 4.

Now, add up the scores for all ten statements. Higher scores on this measure indicate greater psychological flexibility. My score is ___.

Bond, F. W., Hayes, S. C., Baer, R. A., Carpenter, K. M., Orcutt, H. K., Waltz, T. & Zettle, R. D. (2008). Preliminary psychometric properties of the Acceptance and Action Questionnaire—II: A revised measure of psychological flexibility and acceptance. *Manuscript submitted.*

Meaning in Life Questionnaire

The Meaning in Life Questionnaire measures how much you find your life, your very existence, to be important and significant. It provides a numerical score for two subscales, one called the Presence of Meaning and the other called the Search for Meaning. Read each statement and circle the number that most truthfully and accurately describes how you feel about it. Remember that there is no right or wrong answer; whatever you answer is right for you.

1	2	3	4	5	6	7
absolutely untrue	mostly untrue	some-what untrue	can't say true or false	some-what true	mostly true	absolutely true

1.	I understand my life's meaning.	1	2	3	4	5	6	7
2.	I am looking for something that makes my life feel meaningful.	1	2	3	4	5	6	7
3.	I am always looking to find my life's purpose.	1	2	3	4	5	6	7
4.	My life has a clear sense of purpose.	1	2	3	4	5	6	7
5.	I have a good sense of what makes my life meaningful.	1	2	3	4	5	6	7
6.	I have discovered a satisfying life purpose.	1	2	3	4	5	6	7
7.	I am always searching for something that makes my life feel significant.	1	2	3	4	5	6	7
8.	I am seeking a purpose or mission for my life.	1	2	3	4	5	6	7
*9.	My life has no clear purpose.	1	2	3	4	5	6	7
10.	I am searching for meaning in my life.	1	2	3	4	5	6	7

Before calculating your scores on the two subscales, first reverse-score statement 9, "My life has no clear purpose." If you circled a 1, change it to a 7. Change a 2 to a 6, a 3 to a 5, a 6 to a 2, and a 7 to a 1. If you gave yourself a 4, leave it as a 4. Use the new number when doing the addition for the Presence of Meaning score.

The Presence of Meaning subscale measures the degree to which you ascribe meaning to your life right now. To find it, add the numerical values you circled for statements 1, 4, 5, 6, and 9. My Presence of Meaning score is ___.

The Search for Meaning subscale measures the degree to which you consistently search for meaning in your life—exploring, finding, and creating meaning. To find it, add the numerical values you circled for statements 2, 3, 7, 8, and 10. My Search for Meaning score is ___.

Steger, M. F., Frazier, P., Oishi, S., & Kaler, M. (2006). The Meaning in Life Questionnaire: Assessing the presence of and search for meaning in life. *Journal of Counseling Psychology, 53,* 80–93.

Making Sense of Your Curiosity and Well-Being Profile Scores

Now you have determined the degree to which you are a curious explorer (with the Curiosity and Exploration Inventory-II), your basic psychological needs are satisfied (with the Basic Psychological Needs Scale), you work with instead of avoid negative feelings such as anxiety and show signs of psychological flexibility (with the Acceptance and Action Questionnaire-II), and your life is meaningful and you are actively searching for meaning (with the Meaning in Life Questionnaire). To make sense of your scores, it is useful to compare yourself to the "average person." The people you consider your comparison group depend on what is important to you: gender, sexual orientation, race and ethnicity, region of the world in which you are living, occupation, and so on. To keep it simple, I am only going to be focusing on age. If you want to know more about your score, you can find more information in the articles that I list in this book and on my website, www.mason.gmu.edu/tkashdan.

Keep in mind that these measures are not the only tools to measure curiosity and well-being. Rather, they have been shown to be useful in scientific research and are the easiest to use in a book format. The important thing to remember is that no matter how you score on these instruments, using the exercises in this book, you can become more curious and attain greater fulfillment in your life.

Curiosity and Exploration Inventory-II

The average adult completing this questionnaire scores about 17.5 on the Stretching, and 15.5 on the Embracing elements of curiosity. This means that if you score above either of these numbers, you are scoring higher than approximately 50% of other adults. If you score above 33 on your total score, this means that about 50% of other adults are generally less curious than you.

Basic Psychological Needs Scale

If you score above 34 in satisfying Autonomy needs, 30 in satisfying Competence needs, and 44.5 in satisfying Relatedness needs, this means that about 50% of other adults are less satisfied than you.

Acceptance and Action Questionnaire-II

As for psychological flexibility, college students and older adults in the community tend to score higher (averaging a bit below 51) than adults in treatment for emotional difficulties (averaging a bit below 40). If your score is higher than people in these comparison groups, then you are able to contact the present moment and the thoughts and feelings it contains without needless defense, and show greater psychological flexibility in persisting toward valued goals than the "average person."

Meaning in Life Questionnaire

As for meaning in life, age matters much more than the other elements of well-being. When it comes to the presence of meaning in life, teenagers tend to score lower (averaging about 21), college students tend to score higher (averaging a bit below 24), working adults aged 25–44 score a bit lower (averaging close to 22.5), older adults aged 45–64 score higher (averaging close to 25), and older adults aged 65 and up score the highest (averaging a bit below 27). If your score is higher than people in your age group, then you recognize greater meaning in your life than the "average person."

When it comes to searching for meaning in life, people aged 13–44 score the highest (averaging about 25), working adults aged 45–64 score lower (averaging close to 23), and older adults aged 65 and up score the lowest (averaging a bit above 21).

The next two instruments map onto the discussion of values, interests, passions, and purpose in life discussed in Chapters 5 and 9. These two instruments, the Valued Living Questionnaire-II (VLQ-II) and the BULLs-eye Instrument about valued life (BULLI), will help you judge various areas of your life as to what is meaningful. These instruments can help you clarify the most opportune places to begin discovering a purpose.

I suggest that you print out multiple copies of these scales instead of writing in this book. This is because you can use these instruments to track whether your daily actions are aligned with what you care about, what is meaningful to you, and what gives your life purpose. This includes identifying and managing obstacles to purposeful living. Try to do this tracking on a regular, weekly basis.

Valued Living Questionnaire-II (VLQ-II)

The VLQ-II has two parts. Below are ten different areas of life that are valued by some people. I am concerned about your quality of life in each of these areas. In this first part, I want you to ask yourself the following question when you make your ratings for each area.

Current Importance: How important is this area at this time in your life? Rate the importance on a scale of 1-10: 1 means the area is not at all important and 10 means that the area is very important.

Not everyone will value all of these areas, or value all areas the same. Rate each area according to your own personal view of each area. There are no right or wrong answers. Circle the answer that reflects how you feel.

1	2	3	4	5	6	7	8	9	10
not at all important							extremely important		

		1	2	3	4	5	6	7	8	9	10
1.	Family (other than marriage or parenting)	1	2	3	4	5	6	7	8	9	10
2.	Marriage/couples/ intimate relations	1	2	3	4	5	6	7	8	9	10
3.	Parenting	1	2	3	4	5	6	7	8	9	10
4.	Friends/social life	1	2	3	4	5	6	7	8	9	10
5.	Work	1	2	3	4	5	6	7	8	9	10
6.	Education/training	1	2	3	4	5	6	7	8	9	10
7.	Recreation/fun	1	2	3	4	5	6	7	8	9	10
8.	Spirituality	1	2	3	4	5	6	7	8	9	10
9.	Community life	1	2	3	4	5	6	7	8	9	10
10.	Physical self care (diet, exercise, sleep)	1	2	3	4	5	6	7	8	9	10

In the second part, the focus is on how you behave or live. I want you to make several ratings about how consistently you act in ways that map onto your values. Ask yourself the following three questions for each area.

Action: How much have you acted in the service of this area <u>during the past week</u>? Rate your level of action on a scale of 1–10: 1 means you <u>have not been active at all with this value</u> and 10 means you <u>have been very active with this value</u>.

Satisfied with Level of Action: How satisfied are you with your level of action in this area <u>during the past week</u>? Rate your satisfaction with your level of action on a scale of 1–10: 1 means you are <u>not at all satisfied</u> and 10 means you are <u>completely satisfied</u> with your level of action in this area.

Concern: How concerned are you that this area will not progress as you want? Rate your level of concern on a scale of 1–10: 1 means that you are <u>not at all concerned</u> and 10 means that you are <u>very concerned</u>.

Write in an answer, from 1-10, for each question in each area. Do not rate how you lived in the past, how you wish you acted, or how other people would rate you. Rather, know that you will do better in some domains than in others and be honest with yourself.

		Action	Satisfied with Action	Concern
1.	Family (other than marriage or parenting)			
2.	Marriage/couples/ intimate relations			
3.	Parenting			
4.	Friends/social life			
5.	Work			
6.	Education/training			
7.	Recreation/fun			
8.	Spirituality			
9.	Community life			
10.	Physical self care (diet, exercise, sleep)			

Now, compare your answers on the first part of the VLQ-II to the second part of the VLQ-II. How consistent is the way you have lived this past week with the values that you think are important? If you are not living in ways that are aligned with your values, what we might call value-based living, return to Chapters 4, 5 and 9 to infuse your life with more meaning and motivate yourself to devote effort to what you care about. Also, complete the next measure.

You can find more information about this measure from the authors of the VLQ-II:

Wilson, K. G. & DuFrene, T. (2009). *Mindfulness for two: An Acceptance and Commitment Therapy approach to mindfulness in psychotherapy.* Oakland, CA: New Harbinger.

The BULLs-eye Instrument about Valued Life (BULLI)

The BULLI is a tool designed to help people clarify their own values. It is not meant to measure goals, which are things that can be crossed off a list. Goals are things like "I want to get married," "I want a better job," "I want to buy a house," "I want to graduate." Instead, values are always unfinished business and require ongoing action. Values are things like "I want to be the best parent I can be," "I want to be loving, caring, and supportive of my partner," "I want to always give my best at work." Values define for each of us how we want to live, how we choose to interact with other people, and how we behave. Values reflect what we want to do, when we want to do it, and how we want to do it. Understanding our values helps each of us understand how we want to behave in every aspect of life.

There are three parts to the BULLI. The first part will help you clarify your values. The second part is meant to help you identify any obstacles that are keeping you from acting in accordance with your personal values. The third part is to help you create an action plan for bringing the way you live closer to the values you espouse.

Part 1: The Bull's Eye

Part 1 of the BULLI has two exercises. The first exercise asks you to consider four areas, or domains, of your life: work/education, relationships, personal growth/health, and leisure. Read the descriptions following each domain header and write down your values for the area. Think in terms of general life directions and remember that there are no right or wrong answers. Writing down your personal values is what is important about this exercise.

Work/Education: Refers to your workplace and career, education and knowledge, and skills development. (This may include volunteering and other forms of unpaid work.) How do you want to act toward your clients, customers, colleagues, employees, fellow workers? What personal qualities do you want to bring to your work? What skills do you want to develop?

_____

_____

_____

_____

Relationships: Refers to intimacy, closeness, friendship, and bonding in your life. It includes relationships with your partner, children, parents, relatives, friends, coworkers, and other social contacts. What sort of relationships do you want to build? How do you want to be in these relationships? What personal qualities do you want to develop?

_____

_____

_____

_____

Personal Growth/Health: Refers to your ongoing development as a human being. This may include organized religion, personal expressions of spirituality, creativity, developing life skills, meditation, yoga, getting out into nature, exercise, nutrition, and addressing health-risk factors like smoking.

_____

_____

_____

_____

Leisure: Refers to how you play, relax, stimulate, enjoy yourself, and engage in hobbies or other activities for rest, recreation, fun, and creativity.

_____

_____

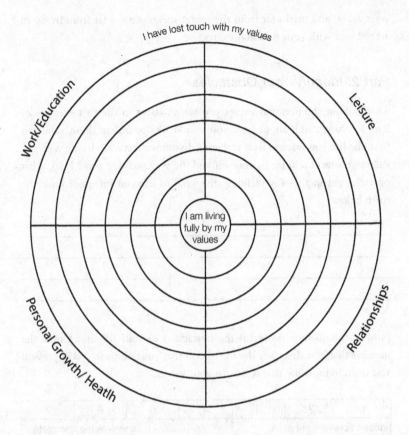

The second exercise in part 1 of the BULLI consists of a Bull's Eye, which looks like a dartboard. Unlike a dartboard, it is further divided into the four domains of life that you just wrote about: work/education, relationships, personal growth/health, and leisure. Read through what you wrote about your values and put an X in each of the four domains to represent how well you live according to your values. Just like a dartboard, the closer to the center the better. Close marks mean you are living fully by your values and marks far from the center mean you are far from living in accordance with your personal values.

Part 2: Identify Your Obstacles

Think about the personal values you wrote about in the first exercise of Part 1. Now, think about how you would like to utilize those values to have the life you want to live. Is there a disconnect between living with the values you most admire in yourself and the way you live now? If so, what obstacles get in the way of living the kind of life you imagine? Describe them below.

_____

_____

_____

_____

Now, estimate how powerful the obstacle(s) in your life are. Circle the number that best describes the likelihood that your obstacle(s) will prevent you from living a life that is true to your values.

1	2	3	4	5	6	7
doesn't prevent me at all					prevents me completely	

Part 3: My Valued Action Plan

So far you have decided what values are most important to you, how close or how far you are from living a life infused with your values, and identified any obstacles that exist that might keep you from achieving a life lived according to your values. Now, think about any actions that you can take that would move you closer to living fully with your values. These actions can be large or small and will be very personal. They might be small movements toward a goal or actions that reflect who you want to be as a person. Use the anxiety that comes with facing an obstacle and deciding to overcome it, or some small part of it, to move closer to living a life fully engaged with your value system. Below, try to write at least one value-based action you are willing to take in each of the four life domains.

Work/education: _____

Relationships: _____

Personal growth/health: _____

Leisure: _____

The BULLs-eye Instrument about valued life was authored by Swedish ACT therapist Tobias Lundgren, modified by Russ Harris, and further modified by myself.

Notes and References

Chapter 1
Seeking a Fulfilling Life:
Why There Is More to It Than Happiness

p. 1   SURVEY OF MORE THAN 10,000 PEOPLE FROM 48
        COUNTRIES: Oishi, S., Diener, E., & Lucas, R. (2007). The
        optimum level of well-being: Can people be too happy? *Perspectives
        on Psychological Science, 2,* 346–360.

p. 2   SOLD ON THE IDEA THAT BEING HAPPY IS THE ONLY
        OR MOST IMPORTANT GOAL IN LIFE: Kashdan, T. B.,
        Biswas-Diener, R., & King, L. A. (2008). Reconsidering happiness:
        The costs of distinguishing between hedonics and eudaimonia.
        Journal of Positive Psychology, 3, 219–233.

p. 2   ENDLESS, FRUITLESS TREADMILL OF "HAPPINESS
        SEEKING": Schooler, J., Ariely, D., & Loewenstein, G. (2003).
        The pursuit and assessment of happiness can be self-defeating. In I.
        Brocas and J. Carrillo (Eds), *Psychology and Economics,* Vol 1
        (pp. 41–70). Oxford, UK: Oxford University Press.; Wilson,
        T. D. & Gilbert, D. T. (2005). Affective forecasting: Knowing what
        to want. *Current Directions in Psychological Science, 14,* 131–134.

p. 3   CURIOSITY IS A DEEPER, MORE COMPLEX PHENOM-
        ENON: Panksepp, J. (1998). Seeking systems and anticipatory states
        of the nervous system. In J. Panksepp (Ed.), *Affective neuroscience: The
        foundations of human and animal emotions* (pp. 144–163). New York:
        Oxford University Press; Silvia, P. J. (2006). *Exploring the psychology
        of interest.* New York: Oxford University Press.

p. 4   THEODORE ROOSEVELT: Morris, E. (1979). *The Rise of Theodore Roosevelt*. New York, NY: Random House; Morris, E. (2001). *Theodore Rex*. New York, NY: Random House.

pp. 6–7   WHY IS CURIOSITY SO IMPORTANT: Kashdan, T. B. (2004). Curiosity. In C. Peterson and M. E. P. Seligman, (Eds), *Character strengths and virtues: A handbook and classification* (pp. 125–141). Washington, DC: American Psychological Association and Oxford University Press.; Kashdan, T. B. & Silvia, P. (2009). Curiosity and interest: The benefits of thriving on novelty and challenge. In S. J. Lopez & C. R. Snyder (2nd Ed.), *Handbook of Positive Psychology*. Oxford, UK: Oxford University Press; Silvia, P. J. (2006), op. cit.

p. 6   IMPORTANCE OF CURIOSITY TO THOUGHT AND MEMORY ARE SO EXTENSIVE: Tomkins, S. S. (1962). *Affect, imagery, consciousness. Vol. 1. The positive affects*. New York: Springer. See p. 347.

p. 10   AN AMAZING ABILITY TO RAPIDLY READJUST OUR MOODS IN RESPONSE TO CHANGES IN OUR CIRCUM-STANCES:" Diener, E., Lucas, R., & Scollon, C. N. (2006). Beyond the hedonic treadmill: Revising the adaptation theory of well-being. *American Psychologist, 61,* 305–314.; Lucas, R. E. (2007). Long-term disability is associated with lasting changes in subjective well-being: Evidence from two nationally representative longitudinal studies. *Journal of Personality and Social Psychology, 92,* 717–730.; Lucas, R. E. (2007). Adaptation and the set-point model of subjective well-being: Does happiness change after major life events? *Current Directions in Psychological Science, 16,* 75–80.; Lucas, R. E., Clark, A. E., Georgellis, Y., & Diener, E. (2003). Re-examining adaptation and the setpoint model of happiness: Reactions to changes in marital status. *Journal of Personality and Social Psychology, 84,* 527–539.; Wilson, T. D., & Gilbert, D. T. (2008). Explaining away: A model of affective adaptation. *Perspectives on Psychological Science, 3,* 370–386.

Chapter 2
The Curiosity Advantage:
Opening the Gateway to What Makes Life Most Worth Living

p. 19   CURIOSITY IS THE ENGINE OF GROWTH: Berlyne, D. E. (1960). *Conflict, arousal, and curiosity.* New York: McGraw-Hill.; Berlyne, D. E. (1971). *Aesthetics and psychobiology.* New York:

Appleton-Century-Crofts.; Dewey, J. (1913). *Interest and effort in education*. Boston: Riverside.; Deci, E. L. (1992). The relation of interest to the motivation of behavior: A self-determination theory perspective. In K. A. Renninger, S. Hidi, & A. Krapp (Eds.), *The role of interest in learning and development* (pp. 43–70). Hillsdale, NJ: Lawrence Erlbaum Associates.; Izard, C. E. (1977). *Human emotions*. New York: Plenum.; Kashdan, T. B. (2004), op. cit.; Kashdan, T. B. & Silvia, P. (2009), op. cit.; Ryan, R. M., & Deci, E. L. (2000). Self-determination theory and the facilitation of intrinsic motivation, social development, and well-being. *American Psychologist, 55,* 68–78.; Silvia, P. J. (2006). *Exploring the psychology of interest*. New York: Oxford University Press.; Spielberger, C. D., & Starr, L. M. (1994). Curiosity and exploratory behavior. In H. F. O'Neil, Jr., & M. Drillings (Eds.) *Motivation: Theory and research* (pp. 221–243). Hillsdale, NJ: Erlbaum.

p. 19   ALL OF US, TO VARYING DEGREES, ARE DRIVEN TO SEEK OUT NEW AND UNCERTAIN EXPERIENCES: Darwin, C. (1965). The expression of emotions in man and animals. Chicago: Chicago University Press. (Original work published 1872.); Ellsworth, P. C. (2003). Confusion, concentration, and other emotions of interest: Commentary on Rozin and Cohen (2003). *Emotion, 3,* 81–85.; Izard, C. A. (1972). *The face of emotion*. New York: Appleton-Century-Crofts.; Langsdorf, P., Izard, C. E., Rayias, M., & Hembree, E. A. (1983). Interest, expression, visual fixation, and heart rate changes in 2- to 8-month-old infants. *Developmental Psychology, 19,* 418–426.; Tomkins, S. S. (1962), op. cit.

p. 20   CURIOSITY ISN'T ALWAYS A JOYFUL PROCESS: Litman, J. A. (2005). Curiosity and the pleasures of learning: Wanting and liking new information. *Cognition and Emotion, 19,* 793–814.; Litman, J. A., & Silvia, P. J. (2006). The latent structure of trait curiosity: Evidence for interest and deprivation curiosity dimensions. *Journal of Personality Assessment, 86,* 318–328.; Loewenstein, G. (1994). The psychology of curiosity: A review and reinterpretation. *Psychological Bulletin, 116,* 75–98.

pp. 21–   TRYING TO MAKE SENSE OF THE UNKNOWN ADDS TO
23   THE INTENSITY OF OUR POSITIVE EXPERIENCES AND MAKES THEM LAST LONGER: Bar-Anan, Y., Wilson, T. D., & Gilbert, D. T. (in press). The feeling of uncertainty intensifies affective reactions. *Emotion*; Berns, G. S., McClure, S. M., Pagnoni, G., &

Montague, P. R. (2001). Predictability modulates human brain response to reward. *Journal of Neuroscience, 21,* 2793–2798.; Knutson, B. & Cooper, J. C. (2006). The lure of the unknown. *Neuron, 51,* 280–282.; Kurtz, J. L., Wilson, T. D., & Gilbert, D. T. (2007). Quantity versus uncertainty: When winning one prize is better than winning two. *Journal of Experimental Social Psychology, 43,* 979–985.; Schultz, W., Dayan, P., & Montague, P. R. (1997). A neural substrate of prediction and reward. *Science, 275,* 1593–1599.; Whitchurch, E., Wilson, T. D., & Gilbert, D. T. (2008). [Not knowing for sure increases positive mood]. Unpublished raw data.

p. 22   OLYMPIC ATHLETES: Sheppard, J. A. & McNulty, J. K. (2002). The affective consequences of expected and unexpected outcomes. *Psychological Science, 13,* 85–88. Note: In a study of Olympic athletes, those entering the competition expecting to win were happier winning a bronze medal than a silver medal. Why? Because the bronze medalists were comparing their win to the possibility of falling out of the medal ranks, whereas silver medalists were comparing themselves to the gold medal they "should" have won.

p. 22–
23   WHY IT IS PARTICULARLY MEANINGFUL WHEN STRANGERS ARE KIND TO US: Wilson, T. D., Centerbar, D. B., Kermer, D. A., & Gilbert, D. T. (2005). The pleasures of uncertainty: Prolonging positive moods in ways people do not anticipate. *Journal of Personality and Social Psychology, 88,* 5–21.

pp. 23–
25   COUNTERBALANCE TO CERTAINTY, CLOSURE, AND CONFIDENCE: For more information on problems with being rigid, intolerance of ambiguity, conformity, and the need for certainty, closure, and structure. Brashers, D. E. (2001). Communication and uncertainty management. *Journal of Communication, 51,* 477–497.; Kruglanski, A. W. (2004). *The psychology of closed mindedness.* New York: Psychology Press.; Kruglanski, A. W., & Webster, D. M. (1996). Motivated closing of the mind: "Seizing" and "freezing." *Psychological Review, 103,* 263–283.; McCrae, R. R. (1996). Social consequences of experiential openness. *Psychological Bulletin, 120,* 323–337.; Sorrentino, R. M. (1995). Reducing self-discrepancies or maintaining self-congruence? Uncertainty orientation, self-regulation, and performance. *Journal of Personality and Social Psychology, 68,* 485–497.; Sorrentino, R. M., Hanna, S. E., Holmes, J. G., & Sharp, A. (1995). Uncertainty orientation and trust in close relationships: Individual differences in cognitive styles. *Journal of Personality and Social Psychology, 68,* 314–327.; Sorrentino,

R. M., Hewitt, E. C., & Raso-Knott, P. A. (1992). Risk taking in games of chance and skill: Informational and affective influences on choice behavior. *Journal of Personality and Social Psychology, 62,* 52–533.; Sorrentino, R. M., & Roney, C. J. R. (2000). *The uncertain mind: Individual differences in facing the unknown.* Philadelphia, PA: Psychology Press.

pp. 27–
28
MINDFULNESS: Brown, K. W., Ryan, R. M., & Creswell, J. D. (2007). Mindfulness: Theoretical foundations and evidence for its salutary effects. *Psychological Inquiry, 18,* 1–26.; Hanh, T. N. (1996). *The miracle of mindfulness: A manual on meditation.* Beacon Press.; Hanh, T. N. (1998). *The heart of Buddha's teaching.* New York: Broadway.; Haskel, P., & Hakeda, Y. (1984). *Bankei zen: Translations from the record of Bankei.* New York, NY: Grove.; Kabat-Zinn, J. (1990). *Full catastrophe living: Using the wisdom of your body and mind to face stress, pain and illness.* New York, NY: Delta.; Kabat-Zinn, J. (1995). *Wherever you go, there you are: Mindfulness meditation in everyday life.* New York: Hyperion.

For evidence that curiosity is distinct from other positive emotions relating to the consumption of pleasurable experiences (such as joy, excitement, and love) in terms of mindful attention. Gable, P. A., & Harmon-Jones, E. (2008). Approach-motivated positive affect reduces breadth of attention. *Psychological Science, 19,* 476–482.

p. 29
CONSENSUS ABOUT HOW TO DEFINE MINDFULNESS: Bishop, S. R., Lau, M., Shapiro, S., Anderson, N., Carlson, L., Segal, Z. V., Abbey, S., Speca, M., Velting, D., & Devins, G. (2004). Mindfulness: A proposed operational definition. *Clinical Psychology: Science and Practice, 11,* 230–241.; Lau, M. A., Bishop, S. R., Segal, Z. V., Buis, T., Anderson, N. D., Carlson, L., Shapiro, S., & Carmody, J. (2006). The Toronto mindfulness scale: Development and validation. *Journal of Clinical Psychology, 62,* 1445–1467.; Siegel, D. J. (2007). *The mindful brain: Reflection and attunement in the cultivation of well-being.* New York: W. W. Norton.

Of all the positive emotions, loving-kindness meditation appears to have the greatest impact on increasing people's daily interest or curiosity. Fredrickson, B. L., Cohn, M. A., Coffey, K. A., Pek, J., & Finkel, S. M. (2008). Open hearts build lives: Positive emotions, induced through loving-kindness meditation, build consequential personal resources. *Journal of Personality and Social Psychology, 95,* 1045–1062.

Curiosity is not an only child; it is part of a family of terms used by writers, scientists, and everyday people making conversation to capture the essence of recognizing, seeking out, and showing a preference for the new. Isolated islands of knowledge have proliferated over the last century, and with a few notable exceptions, researchers and experts neglected to join them together. Just as all of the continents of the world were once interconnected as a single supercontinent named Pangea, before they drifted apart to be separated by vast oceans, so curiosity can be seen as a Pangea of concepts that are all interconnected but are separated in our minds by our fields of study and personal experiences. It's important to clarify concepts that are part of the curiosity family; what makes each of these concepts unique is as important as their shared features. If we grasp the totality of this big, happy family, then for the first time we can begin to understand the depth of scientific discoveries on the benefits of being curious. The curiosity family includes the following siblings:

a. **Interest** is the emotion we feel when we are curious. Interest is one of the select universal emotions that you can find in almost any human anywhere in the world (and plenty of animals as well). After only a few months, infants show a preference for certain sights and sounds in their environments. Describing something as "interesting" is no different than saying "it makes us curious."

b. **Flow** is a term defined by psychologist Mihaly Csikszentmihalyi as "being completely involved in an activity for its own sake. The ego falls away. Time flies. Every action, movement, and thought follows inevitably from the previous one, like playing jazz. Your whole being is involved, and you're using your skills to the utmost." As a state of deep concentration and commitment, flow occurs when we are interested and challenged in an activity such that nothing else seems to matter. Although flow usually occurs when we are working toward a clear goal, the reward stems from the activity itself and not any achievement or accomplishment. Flow can be considered an extreme variant of being curious. see Csikszentmihalyi, M. (1990). *Flow: The psychology of optimal experience*. New York: Harper & Row.

c. **Intrinsic motivation** reflects one of the underlying reasons for doing an activity, and curiosity happens to be the centerpiece. In the words of the distinguished psychologists Richard Ryan and Ed Deci, "Perhaps no single phenomenon reflects the positive potential of human nature as much as intrinsic motivation, the inherent tendency to seek out novelty and challenges, to extend and exercise one's capacities, to explore, and to learn." Quoted from Ryan and Deci (2000), op. cit.

d. **Novelty seeking** is one impetus for our curiosity. Absolute novelty seeking is about engaging in pure newness that has never been experienced before. This is quite rare. More common is relative novelty or the uniqueness of objects, people, or situations because of a change in context or perspective. Even though I might know every beat and lyric to "Hey Joe" by Jimi Hendrix, the tempo might *feel* faster or slower depending on my mood (on the back porch drinking tequila; racing through traffic to get home in time to tuck my kids into bed) and what version is being sung (live at Woodstock; a rendition by a wedding singer). In segments of this book, I will be discussing novelty as one of the most relevant sources in becoming curious.

e. **Openness to experience** is a broad personality trait that captures the breadth, depth, and complexity of a person's inner world. Five basic, universal personality traits discovered by psychologists together can describe a person from almost any culture and one of them is openness (among the others are conscientiousness, extraversion, agreeableness, and neuroticism—an easy way to remember them is the acronym OCEAN). A person open to experiences is intellectually curious, willing to try new activities, aware and receptive of their felt emotions. They are also imaginative, appreciative of music and other forms of creative expression, and skeptical of traditional social, religious, and political values. Thus, openness can be considered a hodge podge of ingredients, and curiosity happens to be a major theme but not the only one. After all, you can be highly curious without showing a strong appreciation of the arts or any interest in politics. That being said, people who are open extract great pleasure from variety and novelty, show a strong desire to explore and seek stimulation, and tend to be independent-minded or psychologically flexible. Our knowledge about curiosity can be gleaned from work on this particular personality trait. See John, O. P., & Srivastava, S. (1999). The Big Five trait taxonomy: History, measurement, and theoretical perspectives. In L. A. Pervin and O. P. John (Eds.), *Handbook of personality: Theory and research* (pp. 102–138). New York: Guilford.; McCrae, R. R., & Costa, P. T. (1997).

Conceptions and correlates of openness to experience. In R. Hogan, J. Johnson, & S. Briggs (Eds.), *Handbook of personality psychology* (pp. 825–847). San Diego: Academic Press.

f. Search for meaning refers to a person's desire and effort to enhance their understanding and comprehend the meaning, significance, and purpose of their life. By definition, searching for meaning involves exploring and investigating, and thus can be considered a targeted form of curiosity. see Frankl, V. E. (1963). *Man's search for meaning: An introduction to logotherapy.* New York: Washington Square Press.; Steger, M. F., Kashdan, T. B., Sullivan, B. A., & Lorentz, D. (2008). Understanding the search for meaning in life: Personality, cognitive style, and the dynamic between seeking and experiencing meaning. *Journal of Personality, 76,* 199–228.

p. 32–34   REPEATED EXPERIENCES CAN CHANGE OUR BRAINS: Otherwise known as personality plasticity. Davidson, R. J., Jackson, D. C. & Kalin, N. H. (2000). Emotion, plasticity, context and regulation: Perspectives from affective neuroscience. *Psychological Bulletin, 126,* 890–906.; McCrae, R. R., Costa, P. T., Jr., de Lima, M. P., Simoes, A., Ostendorf, F., Angleitner, A., Marusic I., Bratko, D., Caprara G. V., Barbaranelli C., Chae J. H., & Piedmont R. L (1999). Age differences in personality across the adult life span: Parallels in five cultures. *Developmental Psychology, 35,* 466–477.; McCrae R. R., Costa P. T., Jr., Terracciano A., Parker W. D., Mills C. J., De Fruyt F., & Mervielde I. (2002). Personality trait development from age 12 to age 18: longitudinal, cross-sectional, and cross-cultural analyses. *Journal of Personality and Social Psychology, 83,* 1456–1468.; Roberts, B. W. & Mroczek, D. (2008). Personality trait change in adulthood. *Current Directions in Psychological Science, 17,* 31–35.

p. 34   ADDITIONAL EVIDENCE FOR PERSONALITY CHANGES THAT ARE RELEVANT TO CURIOSITY RELATED TRAITS: Helson, R., & Roberts, B. W. (1994). Ego development and personality change in adulthood. *Journal of Personality and Social Psychology, 66,* 911–920.; Helson, R., & Srivastava, S. (2001). Three paths of adult development: Conservers, seekers, and achievers. *Journal of Personality and Social Psychology, 80,* 995–1010. Davidson, R. J., Kabat-Zinn, J., Schumacher, J., Rosenkranz, M., Muller, D., Santorelli, S. F., Urbanowski, F., Harrington, A., Bonus, K., & Sheridan, J. F. (2003). Alterations in brain and immune function produced by mindfulness meditation. *Psychosomatic Medicine, 65,*

564–570; Lazar, S. W., Kerr, C. E., Wasserman, R. H., Gray, J. R., Greve, D. N., Treadway, M. T., McGarvey, M., Quinn, B. T., Dusek, J. A., Benson, H., Rauch, S. L., Moore, C. I., & Fischl, B. (2005). Meditation experience is associated with increased cortical thickness. *Neuroreport, 16,* 1893–1897.

p. 34   "IF WE TAKE CARE OF THE MINUTES AND MOMENTS, THE HOURS AND DAYS WILL TAKE CARE OF THEM-SELVES" quoted from p. 213, Goleman, D. (2004). *Destructive emotions: A scientific dialogue with the Dalai Lama.* New York: Bantam Doubleday Dell.

p. 35– 36   HEALTH: Bernheimer, H., Birkmayer, W., Hornykiewicz, O., Jellinger, K., & Seitelberger, F. (1973). Brain dopamine and the syndromes of Parkinson and Huntington. Clinical, morphological and neurochemical correlations. *Journal of Neurological Science, 20,* 415–455.; Chong, H., Riis, J. L., McGinnis, S. M., Williams, D. M., Holcomb, P., & Daffner, K. (2008). To ignore or explore: top down modulation of novelty processing. *Journal of Cognitive Neuroscience, 20,* 1–15.; Daffner, K. R., Rentz, D. M., Scinto, L. F. M., Faust, R., Budson, A. E., & Holcomb, P. J. (2001). Pathophysiology underlying diminished attention to novel events in patients with early AD. *Neurology, 56,* 1377–1383.; Daffner, K. R., Ryan, K. K., Williams, D. M., Budson, A. E., Rentz, D. M., Scinto, L. F. M., & Holcomb, P. J. (2005). Age-related differences in novelty and target processing among cognitively high performing adults. *Neurobiology of Aging, 26,* 1123–1295.; Daffner, K. R., Ryan, K. K., Williams, D. M., Budson, A. E., Rentz, D. M., Wolk, D. A., & Holcomb, P. J. (2006). Increased responsiveness to novelty is associated with successful cognitive aging. *Journal of Cognitive Neuroscience, 18,* 1759–1773.; Daffner, K. R., Ryan, K. K., Williams, D. M., Budson, A. E., Rentz, D. M., Wolk, D. A., & Holcomb, P. J. (2006). Age-related differences in attention to novelty among cognitively high performing adults. Biological Psychology, 72, 67–77.; Daffner, K. R., Scinto, L. F. M., Weintraub, S., Guinessey, J., & Mesulam, M. M. (1994). The impact of aging on curiosity as measured by exploratory eye movements. *Archives of Neurology, 51,* 368–376.; Fritsch, T., Smyth, K. A., Debanne, S. M., Petot, G. J., & Friedland, R. P. (2005). Participation in novelty-seeking leisure activities and Alzheimer's disease. *Journal of Geriatric Psychiatry and Neurology, 18,* 134–141.; Swan, G. E., & Carmelli, D. (1996). Curiosity and mortality in aging adults: A 5-year follow-up of the Western Collaborative Group Study. *Psychology and Aging, 11,*

449–453.; Willis, S. L., Tennstedt, S. L., Marsiske, M., Ball, K., Elias, J., Koepke, K. M., Morris, J. N., Rebok, G. W., Unverzagt, F. W., Stoddard, A. M., & Wright, E. (2006). Long-term effects of cognitive training on everyday functional outcomes in older adults. *Journal of the American Medical Association, 296,* 2805–2814.; Wilson, R. S., de Leon, C. F. M., Barnes, L. L., Schneider, J. A., Bienias, J. L., Evans, D. A., & Bennett, D. A. (2002). Participation in cognitively stimulating activities and risk of incident Alzheimer Disease. *Journal of the American Medical Association, 287,* 742–748.

For a debate on this topic, see the following references: Salthouse, T. (2006). Mental exercise and mental aging: Evaluating the validity of the "use it or lose it" hypothesis. *Perspectives on Psychological Science, 1,* 68–87.; Schooler, C. (2007). Use it—and keep it, longer, probably: A reply to Salthouse (2006). *Perspectives on Psychological Science, 2,* 24–29.; Stine-Morrow, E. A. L. (2007). The dumbledore hypothesis of cognitive aging. *Current Directions in Psychological Science, 16,* 295–299. Note: These scientists do reach a consensus that novel, stimulating activities will improve quality of life, and there is mixed evidence that they enhance cognitive functioning. Also, it appears that the amount of effort devoted to passionate engagement in challenging activities is helpful.

For research on the benefits of curiosity in the nonhuman animal literature. Cavigelli, S. A., & McClintock M. K. (2003). Fear of novelty in infant rats predicts adult corticosterone dynamics and an early death. *Proceedings of the National Academy of Sciences,* 100, 16131–16136.; Cavigelli, S. A., Yee, J. R., & McClintock, M. K. (2006). Infant temperament predicts life span in female rats that develop spontaneous tumors. *Hormones & Behavior,* 50, 454–462.

p. 35   ONE OF EVERY 8 MEMBERS OF THE BABY BOOMER GENERATION: Data on Alzheimer's disease were culled from Kiplinger Retirement Report, December 2006 and *The Philadelphia Inquirer,* April 29, 2007.

pp. 36– INTELLIGENCE: Berg, C. A., & Sternberg, R. J. (1985). Response
37   to novelty: continuity versus discontinuity in the developmental course of intelligence. *Advances in Child Development, 19,* 1–47.; Nair, K. U., & Ramnarayan, S. (2000). Individual differences in need for cognition and complex problem solving. *Journal of Research in Personality, 34,* 305–328.; Raine, A., Reynolds, C., Venables, P. H., and Mednick, S. A. (2002). Stimulation seeking and intelligence: A prospective longitudinal study. *Journal of Personality and Social Psychol-*

ogy, 82, 663–674.; Reiss, S. & Reiss, M. (2004). Curiosity and mental retardation: Beyond IQ. *Mental Retardation, 42,* 77–81.

p. 37   ATTEMPTS TO ENHANCE INTELLIGENCE . . . MORE OFTEN THAN NOT HAVE BEEN COLOSSAL FAILURES: Baumeister, A. A., & Bacharach, V. R. (2000). Early generic educational intervention has no enduring effect on intelligence and does not prevent mental retardation. *The Infant Health and Development Program. Intelligence, 28,* 161–192. Note: This is a difficult literature to read, but in general the benefits of Head Start and other early education (pre-kindergarten) programs have been shown to produce few benefits that can be attributed to the programs and not the quality of schools, child motivation, or other variables. For instance, research shows that following Head Start programs, by the time they are in the second grade, children placed in poor-quality schools perform worse than other students (i.e., the immediate benefits "fade out"). Upon analyzing the extensive costs associated with Head Start, the money allocated can certainly be put to better use with better gains. That is, Head Start fails miserably following a cost-benefit analysis. See McKey, R. H., Condelli, L, Ganson, H., Barrett, B. J., McConkey, C., & Plantz, M. C. (1985). The impact of Head Start on children, families, and communities. (DHHS Publication No. (OHDS) 90–31193). Washington, DC: U.S. Government Printing Office.; Schaefer, S. & Cohen, J. (2000). *Making investments in young children: What the research on early care and education tells us.* National Association of Child Advocates, Washington, D.C.; Zigler, E., & Styfco, S. (2004). *The Head Start debates.* Baltimore, MD: Paul H. Brookes Publishing.

pp. 37–   MEANING AND PURPOSE IN LIFE: Baumeister, R. F. (1991). 38   *Meanings of life.* New York: Guilford Press.; Frankl, V. E. (1963), op. cit.; Hidi, S., & Renninger, K. A. (2006). The four-phase model of interest development. *Educational Psychologist, 41,* 111–127.; Higgins, E. T. (2006). Value from hedonic experience *and* engagement. *Psychological Review, 113,* 439–460.; Kashdan, T. B. & Steger, M. F. (2007). Curiosity and pathways to well-being and meaning in life: Traits, states, and everyday behaviors. *Motivation and Emotion, 31,* 159–173.; McKnight, P. E., & Kashdan, T. B. (2007). Purpose in life as a system that creates and sustains health and well-being: An integrative, testable theory. *Review of General Psychology* (note: authors made an equal contribution); Silvia, P. J. (2001). Interest and interests: The psychology of constructive capriciousness. *Review of General Psychology, 5,* 270–290; Steger, M. F., Kashdan, T. B., et al. (2008), op. cit.; Vallerand, R. J., Blanchard, C. M., Mageau, G. A., Koestner, R., Ratelle, C., Léonard, M., Gagné, M., & Marsolais, J. (2003). Les passions de l'âme: On obsessive and harmonious passion. *Journal of Personality and Social Psychology, 85,* 756–767.

pp. 38–39   WE DON'T OFTEN EQUATE CURIOSITY WITH SOCIAL RELATIONSHIPS: Barnes, S., Brown, K. W., Krusemark, E., Campbell, W. K., & Rogge, R. (2007). The role of mindfulness in romantic relationship satisfaction and responses to relationship stress. *Journal of Marital and Family Therapy, 33,* 482–500; Botwin, M. D., Buss, D. M., & Shackelford, T. K. (1997). Personality and mate preferences: Five factors in mate selection and marital satisfaction. *Journal of Personality, 65,* 106-136; Bouchard, G., Lussier, Y., & Sabourin, S. (1999). Personality and marital adjustment: Utility of the Five-Factor Model of personality. *Journal of Marriage and Family, 61,* 651-660; Burpee, L. C., & Langer. E. J. (2005). Mindfulness and marital satisfaction. *Journal of Adult Development, 12,* 43–51; Donnellan, M. B., Conger, R. D., & Bryant, C. M. (2004). The Big Five and enduring marriages. *Journal of Research in Personality, 38,* 481-504; Kashdan, T. B., & Roberts, J. E. (2004). Trait and state curiosity in the genesis of intimacy: Differentiation from related constructs. *Journal of Social and Clinical Psychology, 23,* 792–816; Kashdan, T. B., & Roberts, J. E. (2006). Affective outcomes and cognitive processes in superficial and intimate interactions: Roles of social anxiety and curiosity. *Journal of Research in Personality, 40,* 140–167; Kashdan, T. B., McKnight, P. E., Sherman, R., Fincham, F. D., & Rose, P. (2008). When being curious breeds intimacy: Taking advantage of intimacy opportunities and transforming boring conversations. *Manuscript submitted*; Lee-Baggley, D., Preece, M., & DeLongis, A. (2005). Coping with interpersonal stress: Role of big five traits. *Journal of Personality, 73,* 1141-1180; Neyer, F. J. & Voigt, D. (2004). Personality and social network effects on romantic relationships: A dyadic approach. *European Journal of Personality, 18,* 279-299; Sneed, C. D., McCrae, R. R., & Funder, D. C. (1998). Lay conceptions of the Five-Factor Model and its indicators. *Personality and Social Psychology Bulletin, 24,* 115-126.

    For a review of work showing how characteristics related to closed-mindedness and the inverse of curiosity are linked to social problems. See Kruglanski (2004), Kruglanski & Webster (1996), McCrae (1996), and Sorrentino & Roney (2000), op. cit.

pp. 40–41   HAPPINESS: Bauer, J. J., McAdams, D. P., & Sakaeda, A. R. (2005). Interpreting the good life: Growth memories in the lives of mature, happy people. *Journal of Personality and Social Psychology, 33,* 203–217.; Brdar, I., & Kashdan, T. B. (200). [Strengths and well-being in Croatia]. Unpublished raw data.; Deci, E. L., & Ryan, R. M. (2000). The "what" and "why" of goal pursuits: Human needs and the self-determination of behavior. *Psychological Inquiry, 11,* 227–268.; Diener, E. (2008, July). *Well-being of planet earth.* Keynote address at the European Positive Psychology Summit, Opatija, Croatia.; Gallagher, M. W., & Lopez, S. J. (2007). Curiosity and well-being. *Journal of Positive Psychology, 2,* 236–248.; Hunter, J. P., & Csikszentmihalyi, M. (2003). The positive psychology of interested adolescents. *Journal of Youth and Adolescence, 32,* 27–35.; Kashdan, T. B., Rose, P., & Fincham, F. D. (2004). Curiosity and exploration: Facilitating positive subjective experiences and personal growth opportunities. *Journal of Personality Assessment, 82,* 291–305.; King, L. A., & Hicks, J. A. (2007). Whatever happened to "what might have been"? Regret, happiness, and maturity. *American Psychologist, 62,* 625–636.; Park, N., Peterson, C., & Seligman, M. E. P. (2004). Strengths of character and well-being. *Journal of Social and Clinical Psychology, 23,* 603–619.; Shimai, S., Otake, K., Park, N., Peterson, C., & Seligman, M. E. P. (2006). Convergence of character strengths in American and Japanese young adults. *Journal of Happiness Studies, 7,* 311–322.

Chapter 3
Our Brains Lust for the New

p. 44   OUR BRAINS ARE HARDWIRED FOR CURIOSITY ALONG WITH . . . WORRY: Baumeister, R. F., & Tice, D. M. (1990). Anxiety and social exclusion. *Journal of Social and Clinical Psychology, 9,* 165–196; Gilbert, P. (2001). Evolution and social anxiety: The role of attraction, social competition, and social hierarchies. *Psychiatric Clinics of North America, 24,* 723–751; Leary, M. R. (2004). The sociometer, self-esteem, and the regulation of interpersonal behavior. In K. D. Vohs, & R. F. Baumeister (Eds.), *Handbook of self-regulation: Research, theory, and applications* (pp. 373–391). New York: Guilford Press.; Panksepp, J. (1998), op. cit.

p. 45   NEGATIVITY BIAS: Baumeister, R. F., Bratslavsky, E., Finkenauer, C., & Vohs, K. D. (2001). Bad is stronger than good. *Review of General Psychology, 5,* 323–370.

p. 46   POSITIVITY OFFSET: Cacioppo, J. T., Gardner, W. L., & Berntson, G. G. (1997). Beyond bipolar conceptualizations and measures: The case of attitudes and evaluative space. *Personality and Social Psychology Review, 1*, 3–25.

p. 47   SQUIRRELS CHOOSE REGULAR PLAYMATES OUTSIDE OF THEIR FAMILY CIRCLE: Warren G. Holmes talk as part of seminar series with the Research Center for Group Dynamics at the University of Michigan in 1999. Panksepp, J. (1998), op. cit.; Panksepp, J., Siviy, S., Normansell, L. (1984). The psychobiology of play: Theoretical and methodological perspectives. *Neuroscience and Behavioral Reviews*, 8, 465–492.

    Note: As an important caveat to this discussion, we can only assume that curiosity developed during the course of evolution. We can't interview our ancestors or watch to see if people who were more curious were really better off than their more fearful or apathetic neighbors. That being said, most researchers are confident that without curiosity and exploration, human beings would not have evolved.

p. 49   MOST COMMONLY DISCUSSED AND MISUNDERSTOOD CHEMICAL IN THE BRAIN LINKED TO HAPPINESS: Ashby, F. G., Isen, A. M., & Turken, A. U. (1999). A neuropsychological theory of affect and its influence on cognition. *Psychological Review, 106*, 529–550.; Berridge, K. C. (2007). The debate over dopamine's role in reward: the case for incentive salience. *Psychopharmacology, 191*, 391–431.

p. 50   SKEPTICAL NEUROSCIENTISTS: Alcaro, A., Huber, R., & Panksepp, J. (2007). Behavioral functions of the mesolimbic dopaminergic system: An affective neuroethological perspective. *Brain Research Reviews, 56*, 283–321.; Depue, R. A. (1996). A neurobiological framework for the structure of personality and emotion: Implications for personality disorders. In J. F. Clarkin & M. F. Lenzenweger (Ed.), *Major theories of personality disorder* (pp. 347–391). New York: Guilford.; Depue, R. A., & Collins, P. F. (1999). Neurobiology of the structure of personality: Dopamine, facilitation of incentive motivation, and extraversion. *Behavioral and Brain Sciences, 22*, 491–569.; Panksepp, J., & Moskal, J. (2008). Dopamine and SEEKING: Subcortical "reward" systems and appetitive urges. In A. Elliot (Ed.), *Handbook of approach and avoidance motivation* (pp. 67–87). New York: Psychology Press.

pp. 51–
53
SEEKING VERSUS HAVING REWARDS: Bunzeck, N., & Duzel, E. (2006). Absolute coding of stimulus novelty in the human substantia nigra/VTA. *Neuron, 51,* 369–379.; Knutson, B., Adams, C. S., Fong, G. W. & Hommer, D. (2001). Anticipation of monetary reward selectively recruits nucleus accumbens. *Journal of Neuroscience, 21,* RC159; Knutson, B., Fong, G. W., Adams, C. S., & Hommer, D. (2001). Dissociation of reward anticipation versus outcome with event-related FMRI. *NeuroReport, 12,* 3683–3687.; Knutson, B. & Peterson, R. (2005). Neurally reconstructing expected utility. *Games and Economic Behavior, 52,* 305–315.; Knutson, B., Wimmer, G. E., Kuhnen, C. M., & Winkielman, P. (2008). Nucleus accumbens activation mediates the influence of reward cues on financial risk taking. *NeuroReport, 19,* 509–513.

pp. 53–
54
SURGE OF OPIATES IN THE BRAIN: Pecina, S. (2008). Opioid reward 'liking' and 'wanting' in the nucleus accumbens. *Physiology and Behavior, 94,* 675–680.; Smith, K. S., & Berridge, K. C. (2007). Opioid limbic circuit for reward: interaction between hedonic hotspots of nucleus accumbens and ventral pallidum. *Journal of Neuroscience, 27,* 1594–1605.

p. 56
HIPPOCAMPUS AND ITS ROLE IN MEMORY: Lisman, J. E., & Grace, A. A. (2005). The hippocampal-VTA loop: controlling the entry of information into long-term memory. *Neuron, 46,* 703–713.; Wittmann, B. C., Bunzeck, N., Dolan, R. J., & Duzel, E. (2007). Anticipation of novelty recruits reward system and hippocampus while promoting recollection. *Neuroimage, 38,* 194–202.

Chapter 4
The Curious Moment:
Sparking Intrigue in the New and Meaning in the Mundane

p. 59
DID THESE OPPORTUNITIES ALLOW US TO SPEND MORE TIME DOING WHAT WE CARED ABOUT MOST: Myers, D. G. (2000). *The American paradox: Spiritual hunger in an age of plenty.* New Haven: Yale University Press.

pp. 59–
60
TRANSFORM MUNDANE, UNSATISFYING TASKS: Sansone, C., & Thoman, D. B. (2005). Interest as the missing motivator in self-regulation. *European Psychologist, 10,* 175–186; Silvia (2006), op. cit.

p. 60   AVERAGE AMERICAN'S DAILY ROUTINE: Kahneman, D., Krueger A. B., Schkade, D., Schwarz, N., & Stone, A. (2004). A survey method for characterizing daily life experience: The day reconstruction method. *Science, 306,* 1776–1780.; Krueger, A. B., Kahneman, D., Schkade, D., Schwarz, N., & Stone, A. (in press). National time accounting: The currency of life. In A. B. Krueger (Ed.), *National time accounting and subjective well-being.* Chicago, IL: University of Chicago Press; Schwarz, N., Kahneman, D., & Xu, J. (in press). Global and episodic reports of hedonic experience. In R. Belli, D. Alwin, & F. Stafford (Eds.), *Using calendar and diary methods in life events research.* Newbury Park, CA: Sage.

p. 63   WORK PROVIDES AN IDEAL OUTLET TO CAPITALIZE ON OUR STRENGTHS: Csikszentmihalyi, M., & LeFevre, J. (1989). Optimal experience in work and leisure. *Journal of Personality and Social Psychology, 56,* 815–822.; Wrzesniewski, A., McCauley, C. R., Rozin, P., & Schwartz, B. (1997). Jobs, careers, and callings: People's relations to their work. *Journal of Research in Personality, 31,* 21–33.

p. 67   OUR HOBBIES, INTERESTS . . . FALL INTO THE CATEGORY OF DESIRABLE ACTIVITIES: Waterman, A. S. (2005). When effort is enjoyed: Two studies of intrinsic motivation for personally salient activities. *Motivation and Emotion, 29,* 165–188.; Fisher, C. D., & Noble, C. S. (2004). A within-person examination of correlates of performance and emotions while working, *Human Performance, 17,* 145–168.

p. 71   AVERAGE HOUSEHOLD HAS THE TELEVISION ON FOR AN INCREDIBLE EIGHT HOURS AND 18 MINUTES OF TV PER DAY: November 21, 2008, Nielsen ratings. This is the highest level of viewer hours since Nielsen began recording behavior in the 1950s.

p. 72   MEN FROM THESE CULTURES LEARN FROM ROLE MODELS: Kashdan, T. B., Mishra, A., Breen, W. E., & Froh, J. J. (in press). Gender differences in gratitude: Examining appraisals, narratives, the willingness to express emotions, and changes in psychological needs. *Journal of Personality.*

p. 73   WALTER REED ARMY MEDICAL CENTER: For more details, see A Breath of Hope: Walter Reed Tries Yoga to Counter PTSD by Eileen Rivers, *The Washington Post,* Tuesday, May 6, 2008.

p. 73   YOU NEED TO CATCH INTEREST BEFORE YOU CAN
         HOLD ON TO IT: Dewey, J. (1913), op. cit.

p. 74   IF I FEEL GOOD THEN I MUST BE DOING SOMETHING
         INTERESTING AND FUN: Clore, G. L., Gasper, K., & Garvin, E.
         (2001). Affect as information. In J. P. Forgas (Ed.), *Handbook of affect
         and social cognition* (pp. 121–144). Mahwah, NJ: Lawrence Erlbaum
         Associates, Inc.; Fredrickson, B. L. (1998). What good are positive
         emotions? *Review of General Psychology, 2,* 300–319.

p. 75   VIEW DIFFICULT SITUATIONS AS CHALLENGES INSTEAD
         OF THREATS: Blascovich, J., Mendes. W. B., Tomaka, J., Salomon,
         K., & Seery, M. (2003). The robust nature of the biopsychosocial
         model challenge and threat: A reply to Wright and Kirby. *Personality
         and Social Psychology Review, 7,* 234–243.; Tomaka, J., Blascovich, J.,
         Kelsey, R. M., & Leitten, C. (1993). Subjective, physiological, and
         behavioral effects of threat and challenge appraisal. *Journal of
         Personality and Social Psychology,* 65, 248–260.

         The story on National Public Radio about an assembly line
         worker in a potato chip factory was briefly mentioned in the
         following article: Sansone, C., Weir, C., Harpster, L., & Morgan, C.
         (1992). Once a boring task always a boring task? Interest as a
         self-regulatory mechanism. *Journal of Personality and Social Psychology,
         63,* 106.

p. 77   WE WORK HARDER AND LONGER, AND ARE MORE
         COMMITTED AND RESPONSIBLE, WHEN WE ARE IN A
         GOOD MOOD: Isen, A. M. & Reeve, J. (2005). The influence of
         positive affect on intrinsic and extrinsic motivation: Facilitating
         enjoyment of play, responsible work behavior, and self-control.
         Motivation and Emotion, 29, 295–323.

pp. 77– CHOOSE THE RIGHT GUIDES AND PARTNERS: Bowlby, J.
   79   (1988). *A secure base: Parent–child attachment and healthy human
         development.* New York: Basic Books.; Elliot, A. J., & Reis, H. T.
         (2003). Attachment and exploration in adulthood. *Journal of Personal-
         ity and Social Psychology,* 85, 317–331.; Feeney, B. C. (2004). A secure
         base: Responsive support of goal strivings and exploration in adult
         intimate relationships. *Journal of Personality and Social Psychology,* 87,
         631–648.

pp.80–   PAY ATTENTION TO THREE NOVEL FEATURES: Data
   81   reported in Langer, E. (2005). *On becoming an artist: Reinventing yourself
         through mindful creativity.* New York: Ballantine.; Langer, E. J., &

Notes and References

Pietrasz, L. (1995). From reference to preference. Unpublished manuscript, Harvard University, Cambridge, MA.

For related studies to the above, see: Langer, E. J., Bashner, R. S., & Chanowitz, B. (1985). Decreasing prejudice by increasing discrimination. *Journal of Personality and Social Psychology. 49*, 113–120.; Langer, E. J., Fiske, S., & Taylor, S. E. (1976). Stigma, staring, and discomfort: A novel-stimulus hypothesis. *Journal of Experimental Social Psychology, 12*, 451–463.

pp. 83–84   ENDINGS AFFECT HOW WE RECALL EVENTS: Redelmeier, D., & Kahneman, D. (1996). Patients' memories of painful medical treatments: Real-time and retrospective evaluations of two minimally invasive procedures. *Pain, 66*, 3–8.

For related studies on the peak-end effect, see: Diener, E., Wirtz, D., & Oishi, S. (2001). End effects of rated life quality: The James Dean effect. *Psychological Science, 12*, 124–128; Fredrickson, B. L. (2000). Extracting meaning from past affective experiences: The importance of peaks, ends, and specific emotions. *Cognition and Emotion, 14*, 577–606.

pp. 85–86   WHEN YOU ARE LEADING AND MOTIVATING OTHERS: Hidi, S., & Rinninger, K. A. (2006), op. cit.

If you think conversations about sexuality, violence, and drugs are inappropriate to get people engaged, there are two things to remember. First, some of the most creative artistic masterpieces address these same themes (see Tolstoy's *Anna Karenina*, Mann's *Death in Venice*, and Bernini's *Apollo and Daphne*).

p. 85   WHEN PEOPLE ENJOY THEMSELVES, THEY ARE MORE CREATIVE, EFFICIENT, PRODUCTIVE, COOPERATIVE, AND ARE BETTER AUTONOMOUS DECISION MAKERS: Estrada, C. A., Isen, A. M., & Young, M. J. (1997). Positive affect facilitates integration of information and decreases anchoring in reasoning among physicians. *Organizational Behavior and Human Decision Processes, 72*, 117–135.; Fredrickson, B. L., & Branigan, C. (2005). Positive emotions broaden the scope of attention and thought-action repertoires. *Cognition and Emotion*, 19, 313–332.; Isen, A. M. (2000). Positive affect and decision making. In M. Lewis & J. Haviland-Jones (Eds.), *Handbook of emotions* (2nd ed., pp. 417–435). New York: Guilford.; Isen, A. M. (2003). Positive affect as a source of human strengths. In L. G. Aspinwall & U. M. Staudinger (Eds.), *A psychology of human strengths: Fundamental questions and future directions for a positive psychology* (pp. 179–195). Washington, DC: American Psychological Association.

pp. 86–87   NOT EVERYONE ENTERS A TASK OR TOPIC AT THE SAME PLACE: Durik, A. M., & Harackiewicz, J. M. (2007). Different strokes for different folks: How personal interest moderates the effects of situational factors on task interest. *Journal of Educational Psychology, 99*, 597-610.

For further information about curiosity interventions, see: Kashdan, T. B., & Fincham, F. D. (2004). Facilitating curiosity: A social and self-regulatory perspective for scientifically based interventions. In P. A. Linley & S. Joseph, (Ed.), *Positive psychology in practice* (pp. 482–503). Hoboken, NJ: Wiley.

Chapter 5
Creating Lasting Interests and Passions

p. 91   MOST OF OUR CURIOUS MOMENTS ARE SHORT-LIVED. Loewenstein, G. (1994), op. cit.

p. 94   STUDY OF MORE THAN 1,000 COLLEGE STUDENTS: Harackiewicz, J. M., Barron, K. E., Tauer, J. M., & Elliot, A. J. (2002). Predicting success in college: A longitudinal study of achievement goals and ability measures as predictors of interest and performance from freshman year through graduation. *Journal of Educational Psychology, 94*, 562-575.

For related studies to the above, see: Church, M. A., Elliot, A. J., & Gable, S. L. (2001). Perceptions of classroom environment, achievement goals, and achievement outcomes. *Journal of Educational Psychology, 93*, 43–54.; Harackiewicz, J. M., Barron, K. E., Tauer, J. M., Carter, S. M., & Elliot, A. J. (2000). Short-term and long-term consequences of achievement goals: Predicting interest and performance over time. *Journal of Educational Psychology, 92*, 316–330.; Harackiewicz, J. M., Durik, A. M., Barron, K. E., Linnenbrink, E. A. & Tauer, J. M. (2008). The role of achievement goals in the development of interest: Reciprocal relations between achievement goals, interest and performance. *Journal of Educational Psychology, 100*, 105–122.

pp. 94–95   NEARLY 200 ATHLETES AT A SUMMER FOOTBALL CAMP: Hulleman, C. S., Durik, A. M., Schweigert, S., & Harackiewicz, J. M. (2008). Task values, achievement goals, and interest: An integrative analysis. *Journal of Educational Psychology, 100*, 398–416.

pp. 96–   OUR VALUES INFLUENCE WHY WE DO THE THINGS WE
100   DO: Miller, W. R., & Rollnick, S. (2002). *Motivational interviewing: Preparing people for change* (2nd ed.). New York: Guilford Press.; Sagiv, L. & Schwartz, S. H. (2000). Value priorities and subjective well-being: Direct relations and congruity effects. *European Journal of Social Psychology, 30,* 177–198.; Schwartz, S. H. (1992). Universals in the content and structure of values: Theoretical advances and empirical tests in 20 countries. In M. P. Zanna (Ed.), *Advances in experimental social psychology* (Vol. 24, pp. 1–65). San Diego: Academic.; Schwartz, S. (1994). Are there universal aspects in the structure and contents of human values? *Journal of Social Issues, 50,* 19–45.; Schwartz, S. (1996). *Value priorities and behavior: Applying a theory of integrated value systems.* In C. Seligman, J. M. Olson, & M. P. Zanna (Eds.), The Ontario symposium: The psychology of values (Vol. 8, pp. 1–24). Mahwah, NJ: Lawrence Erlbaum Associates, Inc.

pp. 100–   FINDING THE RIGHT HOME TO ALIGN YOUR ACTIONS
101   WITH YOUR VALUES: Freitas, A. L., & Higgins, E. T. (2002). Enjoying goal-directed action: The role of regulatory fit. *Psychological Science, 13,* 1–6.; Higgins, E. T. (2006). Value from hedonic experience and engagement. *Psychological Review, 113,* 439–460.; Vaughn, L. A., Baumann, J., & Klemann, C. (2008). Openness to experience and regulatory focus: Evidence of motivation from fit. *Journal of Research in Personality, 42,* 886–894. Note: This notion of fit cannot be explained by how enjoyable a task is or how much of a positive mood people feel; it sparks interest and engagement over and above feeling good.

p. 102   FOR POSITIVE EVENTS TO SOLIDIFY: Diener, E., Lucas, R., & Scollon, C. N. (2006), op. cit.

pp. 103–   CAPITALIZATION: Deci, E. L., Guardia, J., Moller, A., Scheiner,
106   M., & Ryan, R. (2006). On the benefits of giving as well as receiving autonomy support: Mutuality in close friendships. *Personality and Social Psychology Bulletin, 32,* 313–327.; Gable, S. L., Gonzaga, G., & Strachman, A. (2006). Will you be there for me when things go right? Supportive responses to positive event disclosures. *Journal of Personality and Social Psychology, 91,* 904–917.; Gable, S. L., Reis, H. T., Impett, E., & Asher, E. R. (2004). What do you do when things go right? The intrapersonal and interpersonal benefits of sharing positive events. *Journal of Personality and Social Psychology, 87,* 228–245.; Maisel, N. C., Gable, S. L., & Strachman, A. (2008). Responsive behaviors in romantic relationships: What behaviors are helpful

in good times and bad? *Personal Relationships*, 15, 317–338.; Pasupathi, M., & Rich, B. (2005). Inattentive listening undermines self-verification in personal storytelling. *Journal of Personality*, *73*, 1051–1086.; Pasupathi, M., Stallworth, L. M., & Murdoch, K. (1998). How what we tell becomes what we know: Listener effects on speakers' long-term memory for events. *Discourse Processes, 26*, 1–25.; Thoman, D. B., Sansone, C., & Pasupathi, M. (2007). Talking about interest: Exploring the role of social interaction for regulating motivation and the interest experience. *Journal of Happiness Studies, 8,* 335–370.; Zorbas, P., & Kashdan, T. B. (2008, November). Self-expansion in couples: The roles of curiosity, capitalization, and social anxiety. Presented at the Annual Conference of the Association for Behavior and Cognitive Therapies, Orlando, FL.

pp. 107–
108   WORK AS A PASSIONATE CALLING: Vallerand, R. J., & Houlfort, N. (2003). Passion at work: Toward a new conceptualization. In S. W. Gilliland, D. D. Steiner, & D. P. Skarlicki (Eds.), *Emerging perspectives on values in organizations* (pp. 175–204). Greenwich, CT: Information Age Publishing.; Wrzesniewski, A., McCauley, C. R., Rozin, P., & Schwartz, B. (1997), op. cit.

p. 108   GALLUP SURVEY IN MORE THAN 131 COUNTRIES: 2000 annual "Attitudes in the American Workplace VI" Gallup Poll sponsored by the Marlin Company.

p. 109   300 WORKERS IN CANADA: Vallerand, R. J., & Houlfort, N. (2003), op. cit.

pp. 111–
112   DR. RAY FOWLER: Fowler, R. D. (2006). Computers, criminals, an eccentric billionaire, and APA: A brief autobiography. *Journal of Personality Assessment, 87,* 234–438.

pp. 118–
119   IF YOU DON'T ENJOY WHAT YOU ARE DOING, ASK YOURSELF WHETHER YOU CAN CHANGE YOUR PERSPECTIVE: Bianco, A. T., Higgins, E. T., & Klem, A. (2003). How "fun/importance" fit affects performance: Relating implicit theories to instructions. *Personality and Social Psychology Bulletin, 29,* 1091–1103.

p. 119   PHYSICS DAY: 2008 Six Flags Theme Parks, Inc.

p. 119   EMINENT ARTISTS, SCIENTISTS, POLITICIANS, BUSINESS PEOPLE, AND EXPLORERS OF THE TWENTIETH CENTURY: Csikszentmihalyi, M. (1997). *Creativity: Flow and the psychology of discovery and invention.* New York, NY: HarperPerennial. Also see Ludwig, A. (1996). *The price of greatness: Resolving the controversy of creativity and madness.* New York, NY: Guilford.

p. 120 VISUALIZE THE BIG PICTURE AND SHARE IT WITH PEOPLE: Hatfield, E., Cacioppo, J. T., & Rapson, R. L. (1994). *Emotional contagion.* New York: Cambridge University Press.; Wanberg, C. R., & Banas, J. T. (2000). Predictors and outcomes of openness to changes in reorganizing a workplace. *Journal of Applied Psychology, 85,* 132–142.

pp. 120– REWARD PEOPLE BEYOND MONEY: Deci, E. L., Koestner,
121 R., & Ryan, R. M. (1999). A meta-analytic review of experiments examining the effects of extrinsic rewards on intrinsic motivation. *Psychological Bulletin, 125,* 627–668.; Deci, E. L., Koestner, R., & Ryan, R. M. (1999). The undermining effect is a reality after all: Extrinsic rewards, task interest, and self-determination. *Psychological Bulletin, 125,* 692–700.

Chapter 6
The Rewards of Relationships:
Infusing Energy and Passion into Social Interactions

p. 125 THEY'D BEEN MARRIED FOR THREE MONTHS NOW. Boyle, T. C. (2000). *A Friend of the Earth.* New York: Viking.

p. 125 THE OPPOSITE OF LOVE. Elie Wiesel interview in *U.S. News and World Report* (New York, Oct. 27, 1986).

p. 129 CHARACTERISTIC THAT SEPARATES VERY HAPPY PEOPLE FROM THE REST OF SOCIETY: Diener, E., & Seligman, M. E. P. (2002). Very happy people. *Psychological Science, 13,* 80–83.

p. 129 PROSTITUTES IN THE SLUMS OF CALCUTTA: Biswas-Diener, R., & Diener, E. (2006). The subjective well-being of the homeless, and lessons for happiness. *Social Indicators Research, 76,* 185–205.

p. 129 FEELING REJECTED AND HURT BY OTHER PEOPLE: DeWall, C. N., Baumeister, R. F., & Vohs, K. D. (in press). Satiated with belongingness? Effects of acceptance, rejection, and task framing on self-regulatory performance. *Journal of Personality and Social Psychology*; MacDonald, G. & Leary M. R. (2005). Why does social exclusion hurt? The relationship between social and physical pain. *Psychological Bulletin, 131,* 202–223.; Leary, M. R., Springer, C., Negel, L., Ansell, E., & Evans, K. (1998). The causes, phenomenology, and consequences of hurt feelings. *Journal of Personality and Social Psychology, 74,* 1225–1237.; Reis, H. T., Sheldon, K. M., Gable, S. L., Roscoe, R., & Ryan, R. (2000). Daily well being: The role of autonomy, competence, and relatedness. *Personality and Social Psychology Bulletin, 26,* 419–435.

p. 130   THE BEST WAY TO EXPAND AND GROW IS TO ENTER CLOSE RELATIONSHIPS: Aron, A., & Aron, E. N. (1986). *Love as expansion of self: Understanding attraction and satisfaction.* New York: Hemisphere; Aron, A., McLaughlin-Volpe, T., Mashek, D., Lewandowski, G., Wright, S. C., & Aron, E. N. (2004). Including close others in the self. *European Review of Social Psychology, 15,* 101–132.; Knee, C. R., Patrick, H., Victor, N. A., Nanayakkara, A., & Neighbors, C. (2002). Self-determination as growth motivation in romantic relationships. *Personality and Social Psychology Bulletin, 28,* 609–619.

p. 132   AFTER FALLING IN LOVE: Aron, A., Paris, M., & Aron, E. N. (1995). Falling in love: Prospective studies of self-concept change. *Journal of Personality and Social Psychology, 69,* 1102–1112.

p. 132   WHEN MARRIED PARTNERS TALK ABOUT THEIR RELATIONSHIP: Sillars, A., Shellen, W., McIntosh, A., & Pomegrante, M. (1997). Relational characteristics of language: Elaboration and differentiation in marital conversations. *Western Journal of Communication, 61,* 403–422.

p. 132   AS A RESULT OF BEING IN A LASTING FRIENDSHIP WE CONSIDER MANY OF OUR FRIENDS' QUALITIES, INTERESTS, AND VALUES TO BE PART OF US: Mashek, D., Aron, A., & Boncimino, M. (2003). Confusions of self with close others. *Personality and Social Psychology Bulletin, 29,* 382–392.

p. 133   VISIBLE IN THE BRAIN: Aron, A., Fisher, F., Mashek, D. J., Strong, G., Li, H. F., & Brown, L. L. (2005). Reward, motivation and emotion systems associated with early-stage intense romantic love: An fMRI study. *Journal of Neurophysiology, 94,* 327–337.; Aron, A., Whitfield-Gabrieli, S. L., and Lichty, W. (2006). Whole brain correlations: Examining similarity across conditions of overall patterns of neural activation in fMRI. In S. S. Sawilowsky (Ed.),

Real data analysis (pp. 365–369). Charlotte, NC: Information Age Publishing.; Mashek, D., Aron, A., & Fisher, H. (2000) Identifying, evoking and measuring intense feelings of romantic love. *Representative Research in Social Psychology, 24,* 48–55.

Gender differences in friendships. Fehr, B. (1996). *Friendship processes.* Thousand Oaks, CA: Sage.; Reis, H. T. (1998). Gender differences in intimacy and related behaviors: Context and process. In D. J. Canary & K. Dindia (Eds.), *Sex differences and similarities in communication: Critical essays and empirical investigations of sex and gender in interaction* (pp. 203–231). Mahwah, NJ: Lawrence Erlbaum Associates.

p. 134   WHEN THE NEED FOR SAFETY TRUMPS THE NEED TO EXPLORE: Bowlby, J. (1988), op. cit.; Elliot, A. J., & Reis, H. T. (2003), op. cit.; Feeney, B. C. (2004), op. cit.; Feeney, B. C. (2007). The dependency paradox in close relationships: Accepting dependence promotes independence. *Journal of Personality and Social Psychology, 87,* 631–648; Green, J. D., & Campbell, W. K. (2000). Attachment and exploration in adults: Chronic and contextual accessibility. *Personality and Social Psychology Bulletin, 26,* 452–461.; Mikulincer, M. (1997). Adult attachment style and information processing: Individual differences in curiosity and cognitive closure. *Journal of Personality and Social Psychology, 72,* 1217–1230.; Mikulincer, M., & Arad, A. (1999). Attachment working models and cognitive openness in close relationships: A test of the chronic and temporary accessibility effects. *Journal of Personality and Social Psychology, 77,* 710–725.

pp. 135–   BEING A CURIOUS PERSON IS BENEFICIAL IN
138   CREATING HEALTHY, MEANINGFUL RELATIONSHIPS: Barnes, S., Brown, K. W., Krusemark, E., Campbell, W. K., & Rogge, R. (2007), op. cit.; Botwin, M. D., Buss, D. M., & Shackelford, T. K. (1997), op. cit.; Bouchard, G., Lussier, Y., & Sabourin, S. (1999), op. cit.; Burpee, L. C., & Langer. E. J. (2005), op. cit.; Davis, D. (1982). Determinants of responsiveness in dyadic interaction. In W. Ickes & E. S. Knowles (Eds.), *Personality, roles, and social behavior* (pp. 85–139). New York: Springer-Verlag; Donnellan, M. B., Conger, R. D., & Bryant, C. M. (2004), op. cit.; Holland, A. S., & Roisman, G. I. (2008). Big five personality traits and relationship quality: Self-reported, observational, and physiological evidence. *Journal of Social and Personal Relationships, 25,* 811–829; Kashdan, T. B., & Roberts, J. E. (2004), op. cit.; Kashdan, T. B., & Roberts, J. E. (2006),

op. cit.; Kashdan, T. B. et al. (2008), op. cit.; Kelly, A. E., & McKil-lop, K. J. (1996). Consequences of revealing personal secrets. *Psychological Bulletin, 120,* 450–465; McCrae, R. R. (1996). Social consequences of experiential openness. *Psychological Bulletin, 120,* 323–337; Neyer, F. J. & Voigt, D. (2004), op. cit.; Sneed, C. D., McCrae, R. R., & Funder, D. C. (1998), op. cit.; NOTE: Several of these studies found that the benefits of curiosity were not due to other personality traits such as general negative affect, positive affect, sensitivity to reward cues, or social anxiety. Thus there is evidence that there is something uniquely valuable about being curious and open to various experiences.

pp. 138–139   NONVERBAL CUES THAT "LEAK" OUT: Mehrabian, A. (1972). *Nonverbal communication.* Aldine-Atherton, Chicago, Illinois.; Mehrabian, A. (1981). *Silent messages: Implicit communication of emotions and attitudes (2nd ed.).* Wadsworth, Belmont, California.

p. 139   CRUCIAL SKILL OF RECOGNIZING THE EMOTIONS OF OTHER PEOPLE: Matsumoto, D., LeRoux, J., Wilson-Cohn, C., Raroque, J., Kooken, K., Ekman, P., Yrizarry, N., Loewinger, S., Uchida, H., Yee, A., Amo, L., & Goh, A. (2000). A new test to measure emotion recognition ability: Matsumoto and Ekman's Japanese and Caucasian Brief Affect Recognition Test (JACBART). *Journal of Nonverbal Behavior, 24,* 179–209.

pp. 145–148   PEAS IN A POD OR OPPOSITES ATTRACT: Aron, A., & Aron, E. N. (1997). Self-expansion motivation and including other in the self. In W. Ickes (Section Ed.) & S. Duck (Ed.), *Handbook of personal relationships* (2nd ed., Vol. 1, pp. 251–270). London: Wiley.; Aron, A., Steele, J., Kashdan, T. B., & Perez, M. (2006). When similars don't attract: Tests of a prediction from the self-expansion model. *Personal Relationships, 13,* 387–396. For similar studies, see: Goldstein, J. W., & Rosenfeld, H. (1969). Insecurity and preferences for persons similar to oneself. *Journal of Personality, 37,* 253–268.; Grush, J. E., Clore, G. L., & Costin, F. (1975). Dissimilarity and attraction: When difference makes a difference. *Journal of Personality and Social Psychology, 32,* 783–789.; Izard, C. E. (1963). Personality similarity and friendship: A follow-up study. *Journal of Abnormal and Social Psychology, 66,* 598–600; Langer, E. J., Bashner, R. S., & Chanowitz, B. (1985). Decreasing prejudice by increasing discrimination. *Journal of Personality and Social Psychology, 49,* 113–120.

pp. 147–148   COMPLEMENTARY APPROACHES TO WORKING TOWARD GOALS: Lake, V. K. B., Lucas G., Molden D. C., Finkel E. J., Coolsen, M. K., Kumashiro M., Rusbult C. E., & Higgins E. T. (in press). When opposites fit: Increased relationship strength from partner complementarity in regulatory focus. *Journal of Personality and Social Psychology.*

p. 150   LONELINESS: Cacioppo, J. T., & Patrick, B. (2008). *Loneliness: Human nature and the need for social connection.* New York: W. W. Norton and Company.; Fredrickson, B. L., & Carstensen, L. L. (1990). Choosing social partners: How old age and anticipated endings make people more selective. *Psychology and Aging, 5,* 335–347.; Iecovich, E. (2004) Social support networks and loneliness among elderly Jews in Russia and Ukraine. *Journal of Marriage and Family* 66, 306–317.; Kessler, R. C., Chiu, W. T., Demler, O., & Walters, E. E. (2005). Prevalence, severity, and comorbidity of 12-month DSM-IV disorders in the national comorbidity survey replication. *Archives of General Psychiatry, 62,* 617–627.; Kessler, R. C., McGonagle, K. A., Zhao, S., Nelson, C. B., Hughes, M., Eshelman, S., et al. (1994). Lifetime and 12-month prevalence of DSM-III-R psychiatric disorders in the United States. *Archives of General Psychiatry, 51,* 8–19.; Killeen, C. (1998). Loneliness: an epidemic in modern society. *Journal of Advanced Nursing, 28,* 762–770.; McCamish-Svensson, Samuelsson, G., & Hagberg, B. (2001). Correlates and prevalence of loneliness from young old to oldest old: Results from a Swedish cohort. *International Journal of Aging, 3,* 1–24.

pp. 151–152   BOREDOM: Leary, M. R., Rogers, P. A., Canfield, R. W., & Coe, C. (1986). Boredom in interpersonal encounters: Antecedents and social implications. *Journal of Personality and Social Psychology, 51,* 968–975.; Lewandowski, G. W., Jr., & Ackerman, R. A. (2006). Something's missing: Need fulfillment and self-expansion as predictors of susceptibility to infidelity. *Journal of Social Psychology, 146,* 389–403.

p. 152   SEVEN-YEAR ITCH: Lucas, R. E. (2007), op. cit.

pp. 153–154   ROMANTIC COUPLES CALL IT QUITS: Center for Disease Control's National Center for Health Statistics 2002.

pp. 154–156   WORKING WITH NEGATIVITY: Fredrickson, B. L., & Losada, M. F. (2005). Positive affect and the complex dynamics of human flourishing. *American Psychologist, 60,* 678–686.; Gottman, J. M. (1993). The roles of conflict engagement, escalation, and avoidance in marital interaction: A longitudinal view of five types of couples.

Journal of Consulting and Clinical Psychology, 61, 6–15.; Gottman, J. M., Coan, J. A., Carrere, S., & Swanson, C. (1998). Predicting marital happiness and stability from newlywed interactions. *Journal of Marriage and Family, 60,* 5–22.; Wilson, K. G. & DuFrene, T. (2009). *Mindfulness for two. An Acceptance and Commitment Therapy approach to mindfulness in psychotherapy.* Oakland, CA: New Harbinger.

pp. 155– PEOPLE WHO ARE CURIOUS AND OPEN TO VARIOUS
156   EXPERIENCES SHOW A GREATER TOLERANCE FOR
HANDLING NEGATIVITY IN SOCIAL INTERACTIONS
AND RELATIONSHIPS: Bollmer, J. M., Harris, M. J., Milich, R., & Georgesen, J. C. (2003). Taking offense: Effects of personality and teasing history on behavioral and emotional reactions to teasing. *Journal of Personality, 71,* 557–603; Bouchard, G., Lussier, Y., & Sabourin, S. (1999), op. cit.; Kashdan, T. B., et al. (2008), op. cit.; Lee-Baggley, D., Preece, M., & DeLongis, A. (2005), op. cit.

p. 156   PEOPLE WHO SPEND THE MOST FACE-TO-FACE TIME
WITH THEIR PARTNERS: Kilbourne, B. S., Howell, F., & England, P. (1990). A measurement model for subjective marital solidarity: Invariance across time, gender, and life cycle stage. *Social Science Research, 19,* 62–81.; Kingston, P. W., & Nock, S. L. (1987). Time together among dual-earner couples. *American Sociological Review, 52,* 391–400.; Orden, S. R., & Bradburn, N. M. (1968). Dimensions of marriage happiness. *American Journal of Sociology, 73,* 715–731.; Orthner, D. K. (1975). Leisure activity patterns and marital satisfaction over the marital career. *Journal of Marriage and the Family, 37,* 91–101.; White, L. K. (1983). Determinants of espousal interaction: marital structure or marital happiness. *Journal of Marriage and the Family, 45,* 511–519.

p. 158   RELATIONSHIP-ENHANCING EXERCISES: Reissman, C, Aron, A., & Bergen, M. R. (1993). Shared activities and marital satisfaction: Causal direction and self-expansion versus boredom. *Journal of Social and Personal Relationships, 10,* 243–254.

pp. 159– WHY TAKING PART IN EXCITING ACTIVITIES WAS
160   LINKED TO GREATER RELATIONSHIP SATISFACTION: Aron, A., Norman, C. C., Aron, E. N., McKenna, C., & Heyman, R. (2000). Couples shared participation in novel and arousing activities and experienced relationship quality. *Journal of Personality and Social Psychology, 78,* 273–283.; Strong, G., & Aron, A. (2006). The effect of shared participation in novel and challenging activities on experienced

relationship quality: Is it mediated by high positive affect? In K. D. Vohs, & E. J. Finkel (Eds.), *Self and relationships: Connecting intrapersonal and interpersonal processes* (pp. 342–359). New York: Guilford Press.

    For related work, see: Aron, A., Norman, C., Aron, E. N., & Lewandowski, G. W., Jr. (2003). Shared participation in self-expanding activities: Positive effects on experienced marital quality. In P. Noller & J. Feeney (Eds.), *Marital interaction* (pp. 177–196). Cambridge University Press.; Graham, J. M. (2008). Self-expansion and flow in couples' momentary experiences: An experience sampling study. *Journal of Personality and Social Psychology, 95,* 679–694.; Lewandowski, G. W., Jr., & Aron, A. (2004). Distinguishing arousal from novelty and challenge in initial romantic attraction between strangers. *Journal of Social Behavior and Personality, 32,* 361–372.; Lewandowski, G. W., Jr., & Aron, A. (2007). The effects of novel/challenging versus arousing activities on couples' experienced relationship quality. *Manuscript submitted;* Lewandowski, G. W., Jr., Aron, A., Bassis, S., & Kunak, J. (2006). Losing a self-expanding relationship: Implications for the self-concept. *Personal Relationships, 13,* 317–331.

pp. 160–
161   MINDFULNESS THERAPY FOR COUPLES: Carson, J. W., Carson, K. M., Gil, K. M., & Baucom, D. H. (2004). Mindfulness-Based Relationship Enhancement. *Behavior Therapy, 35,* 471–494.; Carson, J. W., Carson, K. M., Gil, K. M., & Baucom, D. H. (2007). Self-expansion as a mediator of relationship improvements in a mindfulness intervention. *Journal of Marital and Family Therapy, 35,* 517–528.

pp. 161–
164   BRINGING OTHERS INTO THE MIX: Slatcher, R. B. (2008, January). *Effects of couple friendships on relationship closeness.* Presented at the annual meeting of the Society for Personality and Social Psychology, Albuquerque, NM.

Chapter 7
The Anxious Mind and the Curious Spirit

p. 170   BETH COMSTOCK: Excerpt from *Success Magazine,* May 2008.

pp. 170–
171   YOUNG CHILDREN ARE OFTEN SCARED OF GOING TO THE HOSPITAL: McGrath, P., & Huff, N. (2001). 'What is it?': Findings on preschoolers' response to play with medical equipment. *Child: Care, Health, and Development, 27,* 451–462. Also, see: Abbott, K.

(1990) Therapeutic use of play in the psychological preparation of preschool children undergoing cardiac surgery. *Issues in Comprehensive Pediatric Nursing, 13,* 265–277.; Zahr, L. K. (1998) Therapeutic play for hospitalized preschoolers in Lebanon. *Pediatric Nursing, 23,* 449–454.; Zeigler, D. B. & Prior, M. M. (1994) Preparation for surgery and adjustment to hospitalization. *Nursing Clinics of North America, 29,* 655–669.

Other studies showing that curiosity is an antidote to anxiety. Axtell, C., Wall, T., Stride, C., Pepper, K., Clegg, C., Gardner, P., & Bolden, R. (2002). Familiarity breeds contempt: The impact of exposure to change on employee openness and well-being. *Journal of Occupational and Organizational Psychology, 75,* 217–231.; Beckmann, J., & Trudewind, C. (1997). A functional-analytic perspective on affect and motivation. *Polish Psychological Bulletin, 28,* 125–143.; Chamorro-Permuzic, T., & Reichenbacher, L. (2008). Effects of personality and threat of evaluation on divergent and convergent thinking. *Journal of Research in Personality, 42,* 1095–1101. Trudewind, C. (2000). Curiosity and anxiety as motivational determinants of cognitive development. In J. Heckhausen (Ed.), *Motivational psychology of human development: Developing motivation and motivating development* (pp. 15–38). New York: Elsevier.; Trudewind, C., Mackowiak, K., & Schneider, K. (1999). Neugier, Angst und kognitive Entwicklung [Curiosity, anxiety, and cognitive development]. In Jerusalem, M. & R. Pekrun (Eds.), *Emotion, motivation und leistung* (pp. 105–126). Göttingen, Germany: Hogrefe.; Trudewind, C., Schubert, U., & Ballin, U. (1996). The role of curiosity and anxiety as basic motives for learning in young children. In C. Spiel, U. Kastner-Koller, & P. Deimann (Eds.), Motivation und lernen aus der perspektive lebenslanger entwicklung (pp. 15–30). Munster, Germany: Waxmann.; van Dijk, E., & Zeelenberg, M. (2007). When curiosity killed regret: Avoiding or seeking the unknown in decision-making under uncertainty. *Journal of Experimental Social Psychology, 43,* 656–662.

pp. 172–   ANXIETY AND CURIOSITY SYSTEMS AS TWO KNOBS:
174   Paraphrased from, Hayes, S. C., Strosahl, K., & Wilson, K. G.
           (1999). *Acceptance and Commitment Therapy: An experiential approach to
           behavior change.* New York: Guilford Press.

p. 175   SPEAKERS CHANGE THEIR MINDSET OF WHAT CONSTI-
           TUTES A MISTAKE: Data are reported in: Langer, E. (2005), op. cit.

pp. 176–   OUR SOCIAL ROLES AND IDENTITY: Bardi, A., & Ryff,
177   C. D. (2007). Interactive effects of traits on adjustment to a life
            transition. *Journal of Personality, 75,* 955–984.; Whitbourne, S. K.
            (1986). Openness to experience, identity flexibility, and life change
            in adults. *Journal of Personality and Social Psychology, 50,* 163–168. For
            the problems associated with a need for certainty and closure: Krug-
            lanski, A. W., & Webster, D. M. (1996), op. cit.; Kruglanski, A. W.,
            Pierro, A., Mannetti, L., & De Grada, E. (2006). Groups as epistemic
            providers: Need for closure and the unfolding of group-centrism.
            Psychological Review, 113, 84–100.; Linville, P. W. (1985). Self-
            complexity and affective extremity: Don't put all your eggs in one
            cognitive basket. *Social Cognition,* 3, 94–120.; Rafaeli-Mor, E. &
            Steinberg, J. (2002). Self-complexity and well-being: A research
            synthesis. *Personality and Social Psychology Review,* 6, 31-58.; Roccas,
            S., & Brewer, M. B. (2002). Social identity complexity. *Personality
            and Social Psychology Review, 6,* 88–106.

            For related work on flexible identities and curiosity, see:
            Letzring, T. D., Block, J., & Funder, D. C. (2005). Ego-control and
            ego-resiliency: Generalization of self-report scales based on person-
            ality descriptions from acquaintances, clinicians, and the self. *Journal
            of Research in Personality, 39,* 395–422.

            Curiosity aids in dealing with complexity and ambiguity in
            understanding the self. Adler, J. M., Wagner, J. W., & McAdams, D. P.
            (2007). Personality and the coherence of psychotherapy narratives.
            Journal of Research in Personality, 41, 1179–1198.; Bauer, J. J.,
            McAdams, D. P., & Sakaeda, A. R. (2005). Interpreting the good
            life: Growth memories in the lives of mature, happy people. *Journal
            of Personality and Social Psychology, 33,* 203–217.; McAdams, D. P.,
            Anyidoho, N. A., Brown, C., Huang, Y. T., Kaplan, B., &
            Machado, M. A. (2004). Traits and stories: Links between disposi-
            tional and narrative features of personality. *Journal of Personality, 72,*
            761–782.

pp. 178–   WORK THROUGH THE CONFLICTS AND UNCERTAIN-
179   TIES IN OUR RELATIONSHIPS: Aron, A., Norman, C. C.,
            Aron, E. N., McKenna, C., & Heyman, R. (2000), op. cit.;
            Bollmer, J. M., Harris, M. J., Milich, R., & Georgesen, J. C. (2003),
            op. cit.; Bouchard, G., Lussier, Y., & Sabourin, S. (1999), op. cit.;
            Lee-Baggley, D., Preece, M., & DeLongis, A. (2005), op. cit.; Sorren-
            tino, R. M. (1995), op. cit.; Sorrentino, R. M., Hanna, S. E.,
            Holmes, J. G., Sharp, A. (1995), op. cit.; Sorrentino, R. M.,
            Hewitt, E. C., Raso-Knott, P. A. (1992), op. cit.

pp. 179–
181
WHEN NORMAL FEELINGS OF ANXIETY TURN INTO A PROBLEM: Borkovec, T. D., Alcaine, O., & Behar, E. (2004). Avoidance theory of worry and generalized anxiety disorder. In R. G. Heimberg, C. L. Turk, & D. S. Mennin (Eds.), *Generalized anxiety disorder: Advances in research and practice* (pp. 77–108). New York: Guilford Press.; Eysenck, M. W., Santos, R., Derakshan, N., & Calvo, M. G. (2007). Anxiety and cognitive performance: Attentional control theory. *Emotion, 7,* 3336–3353.; Hayes, S. C., Luoma, J., Bond, F., Masuda, A., & Lillis, J. (2006). Acceptance and Commitment Therapy: Model, processes, and outcomes. *Behaviour Research and Therapy, 1,* 1–25.; Mennin, D. S., Heimberg, R. G., Fresco, D. M., & Turk, C. L. (2005). Preliminary evidence for an emotion dysregulation model of generalized anxiety disorder. *Behaviour Research and Therapy, 43,* 1281–1310.

pp. 181–
182
BEING PHYSICALLY TENSE AND AROUSED IS NOT THE PROBLEM: Kleine, D. (1990). Anxiety and sport performance: A meta-analysis. *Anxiety Research, 2,* 113–131.

pp. 182–
183
PHYSICALLY AND EMOTIONALLY DRAINED: Muraven, M. R., & Baumeister, R. F. (2000). Self-regulation and depletion of limited resources: Does self-control resemble a muscle? *Psychological Bulletin, 126,* 247–259.; Rawn, C. D., & Vohs, K. D. (2006). The importance of self-regulation for interpersonal functioning. In K. D. Vohs & E. J. Finkel (Eds.), *Self and relationships: Connecting intrapersonal and interpersonal processes* (pp. 15–31). New York: Guilford Press.

pp. 183–
186
WITH TOO MUCH ANXIETY, CHAOS AND STAGNATION REIGN: Eifert, G. H., & Forsyth, J. P. (2005). Acceptance and Commitment Therapy for anxiety disorders: A practitioner's treatment guide using mindfulness, acceptance, and values-based behavior change strategies. Oakland, CA: New Harbinger.; Kashdan, T. B. (2007). Social anxiety spectrum and diminished positive experiences: Theoretical synthesis and meta-analysis. *Clinical Psychology Review, 27,* 348–365.; Kashdan, T. B., Elhai, J. D., & Breen, W. E. (2008). Social anxiety and disinhibition: An analysis of curiosity and social rank appraisals, approach-avoidance conflicts, and disruptive risk-taking behavior. *Journal of Anxiety Disorders, 22,* 925–939. Kashdan, T. B., Zvolensky, M. J., & McLeish, A. C. (2008). Anxiety Sensitivity and Affect Regulatory Strategies: Individual and interactive risk factors for anxiety-related symptoms. *Journal of Anxiety Disorders, 22,* 429–440.; Kessler, R. C., et al., (2005), op. cit.

p. 185   MOST SOCIALLY ANXIOUS PEOPLE NEVER SEEK TREATMENT: Olfson, M., Guardino, M., Struening, E., Schneier, F. R., Hellman, F., & Klein, D. F. (2000). Barriers to the treatment of social anxiety. *American Journal of Psychiatry, 157,* 521–527.

p. 186   PSYCHOLOGICAL FLEXIBILITY: Kashdan, T. B. (2007), op. cit.; Kashdan, T. B., & Breen, W. E. (2008). Social anxiety and positive emotions: A prospective examination of a self-regulatory model with tendencies to suppress or express emotions as a moderating variable. *Behavior Therapy, 39,* 1–12; Kashdan, T. B. & Steger, M. F. (2006). Expanding the topography of social anxiety: An experience sampling assessment of positive emotions and events, and emotion suppression. *Psychological Science,* 17, 120–128.

p. 187   SUFFERING FROM ACTIONS NOT TAKEN IS MORE INTENSE AND ENDURING THAN TAKING RISKS AND FAILING: Gilovich, T., Medvec, V. H., & Kahneman, D. (1998). Varieties of regret: A debate and partial resolution. *Psychological Review, 105,* 602–605.

pp. 187–   ELIXIR OF OPTIMAL ANXIETY: Berlyne, D. E. (1960).
188   *Conflict, arousal, and curiosity.* New York: McGraw-Hill; Csikszentmihalyi, M. (1990), op. cit.; Silvia, P. J. (2006), op. cit.; Spielberger, C. D., & Starr, L. M. (1994), op. cit.

p. 189   NOVELTY POTENTIAL AND COPING POTENTIAL: Shepperd, J. A., & McNulty, J. (2002). The affective consequences of expected and unexpected outcomes. *Psychological Sciences, 13,* 84–87; Silvia, P. J. (2005). What is interesting? Exploring the appraisal structure of interest. Emotion, 5, 89–102.; Silvia, P. J. (2008). Appraisal components and emotion traits: Examining the appraisal basis of trait curiosity. *Cognition and Emotion, 22,* 94–113.

p. 190   TITLES OFFER AN ENTRY POINT INTO THE MIND OF THE ARTIST: Millis, K. (2001). Making meaning brings pleasure: The influence of titles on aesthetic experiences. *Emotion, 1,* 320–329.; Silvia, P. J. (2006). Artistic training and interest in visual art: Applying the appraisal theory of aesthetic emotions. *Empirical Studies of the Arts, 24,* 139–161.

pp. 191–   SYNESTHESIA: Dixon, M. J., Smilek, D., Cudahy, C., & Merikle,
192   P. M. (2000). Five plus two equals yellow. Nature, 406, 365.; Cytowic, R. E. (1993). *The man who tasted shapes.* New York:

Putnam.; Sacks, O. (1985). *The Man Who Mistook His Wife for a Hat and Other Clinical Tales*. Summit Books.

pp. 192–
193
SITUATIONS THAT PROVIDE THE OPTIMAL EXCITED SENSE OF BUTTERFLIES: Csikszentmihalyi, M., & LeFevre, J. (1989), op. cit.; Kashdan, T. B. & Yuen, M. (2007). Whether highly curious students thrive academically depends on the learning environment of their school: A study of Hong Kong adolescents. *Motivation and Emotion, 31,* 260–270.; Peters, R. A. (1978). Effects of anxiety, curiosity, and perceived instructor threat on student verbal behavior in the college classroom. *Journal of Educational Psychology, 70,* 388–395. For related work, see: Schiefele, U., Krapp, A., & Winteler, A. (1992). Interest as a predictor of academic achievement: A meta analysis of research. In K. A. Renninger, S. Hidi, & A. Krapp (Eds.), *The role of interest in learning and development* (pp. 183–212). Hillsdale, NJ: Erlbaum.; Smith, J. L., Sansone, C., & White, P. H. (2007). The stereotyped task engagement process: The role of interest and achievement motivation. *Journal of Educational Psychology, 99,* 99–114.; Thoman, D. B. & Sansone, C. (2007). Getting different feedback for similar work: How benefiting from or being hurt by biased feedback affects activity interest, self-esteem, and future motivation. *Manuscript submitted.*

It has been well-established by decades of research and clinical work that the most effective tool for helping people work with their anxiety happens to be exposure-based exercises. This is the active ingredient that makes cognitive-behavioral treatments for anxiety effective. To read more about these strategies, see: Barlow, D. H. (2002). *Anxiety and its disorders: The nature and treatment of anxiety and panic* (2nd ed.). New York: Guilford Press; Wells, A. (2000). Emotional disorders and metacognition: Innovative cognitive therapy. Chichester, UK: Wiley.

pp. 195–
196
A SURVIVAL GUIDE FOR QUICKSAND: Hayes, S. C., & Smith, S. (2005). *Get out of your mind and into your life: The new acceptance and commitment therapy.* Oakland, CA: New Harbinger Publications, Inc.

pp. 196–
197
WARM CHOCOLATE CAKE: from p.124, Hayes, S. C., Strosahl, K., & Wilson, K. G. (1999). *Acceptance and Commitment Therapy: An experiential approach to behavior change.* New York: Guilford Press.

For related work, see: Wegner, D. M. (1994). Ironic processes of mental control. *Psychological Review, 101,* 34–52.

pp. 199– DEFUSING NASTY, FRIGHTENING THOUGHTS:
201   Marcks, B. A. & Woods, D. W. (2005). A comparison of thought
      suppression to an acceptance-based technique in the management of
      personal intrusive thoughts: A controlled evaluation. *Behaviour
      Research and Therapy, 43,* 433–445.; Marcks, B. A. & Woods, D. W.
      (2007). Role of thought-related beliefs and coping strategies in the
      escalation of intrusive thoughts: An analog to obsessive-compulsive
      disorder. *Behaviour Research and Therapy, 45,* 2640–2651.; Masuda, A.,
      Hayes, S. C., Sackett, C. F., & Twohig, M. P. (2004). Cognitive
      defusion and self-relevant negative thoughts: Examining the impact of
      a ninety-year-old technique. *Behaviour Research and Therapy, 42,*
      477–485.

pp. 203– CHOOSING A DIRECTION FOR YOUR LIFE WITH A
204   WELL-CALIBRATED COMPASS: Wilson, K. G. & DuFrene, T.
      (2009). *Mindfulness for Two: An Acceptance and Commitment Therapy
      Approach to Mindfulness in Psychotherapy.* Oakland, CA: New Harbin-
      ger; Wilson, K. G. & Murrell, A. R. (2004). Values work in Accep-
      tance and Commitment Therapy: Setting a course for behavioral
      treatment. In S. C. Hayes, V. M. Follette, & M. Linehan, (Eds.),
      The new behavior therapies: Expanding the cognitive behavioral tradition
      (pp. 120-151). New York: Guilford Press.

Chapter 8
The Dark Side of Curiosity:
Obsessions, Sensational Thrills, Sex,
Death, and Detrimental Gossip

pp. 210– DONKEY KONG: The saga is captured in the 2007 documentary,
211   *The King of Kong: A Fistful of Quarters.*

pp. 212– VACUUMS: *The Tonight Show with Jay Leno.* Episode dated 9
213   February 2007; *CBS Evening News*-Assignment America, "Vacuum
      Boy"; July 27, 2007.

pp. 213– OBSESSIVE INTERESTS: Mageau, G. A., & Vallerand, R. J.
214   (2007). The moderating effect of passion on the relation between
      activity engagement and positive affect. *Motivation and Emotion, 31,*
      312–321.; Rip, B., Fortin, S., & Vallerand, R. J. (2006). The rela-
      tionship between passion and injury in dance students. *Journal of Dance
      Medicine and Science, 10,* 14–20.; Vallerand, R. J., Blanchard, C. M.,
      Mageau, G. A., Koestner, R., Ratelle, C., Léonard, M., Gagné, M., &

Marsolais, J. (2003). Les passions de l'âme: On obsessive and harmonious passion. *Journal of Personality and Social Psychology, 85,* 756–767. ; Vallerand, R. J., Salvy, S. J., Mageau, G. A., Denis, P., Grouzet, F. M. E., & Blanchard, C. B. (2007). On the role of passion in performance. *Journal of Personality, 75,* 505–533.

p. 215 UNREQUITED LOVE. Aron, A., & Aron, E. N. (1997), op. cit.; Aron, A., Aron, E. N., & Allen, J. (1998). Motivations for unreciprocated love. *Personality and Social Psychology Bulletin, 24,* 787–796.

pp. 215–216 DELUSIONAL FANS: Dietz, P. Matthews, D., Van Duyne, C., Martell, D., Parry, C., Stewart, T., Warren, J., & Crowder, J. (1991). Threatening and otherwise inappropriate letters to Hollywood celebrities. *Journal of Forensic Sciences, 36,* 185–209.

pp. 216–219 SOCIAL INFORMATION SEEKING: Jaeger, M. E., Skleder, A. A., Rind, B., & Rosnow, R. L. (1994). Gossip, gossipers, and gossipees. In R. F. Goodman & A. Ben-Ze'ev (Eds.), *Good Gossip* (pp.154–168). Lawrence, KS: Kansas University Press.; Litman, J. A., & Pezzo, M. V. (2005). Individual differences in the attitudes towards gossip. *Personality and Individual Differences, 38,* 963–980.; McAndrew, F. T., Bell, E. K., & Garcia, C. M. (2007). Who do we tell and whom do we tell on? Gossip as a strategy for status enhancement. *Journal of Applied Social Psychology, 37,* 1562–1577.; Wert, S. R., & Salovey, P. (2004). A social comparison account of gossip. *Review of General Psychology, 8,* 122–137.

pp. 219–220 TED BUNDY: Newton, M. (2006). *The Encyclopedia of Serial Killers* (2nd ed.). New York: Checkmark Books.; Vronsky, P. (2004). *Serial killers: The method and madness of monsters* (pp. 102–142). New York: The Berkley Publishing Group.; *New York Times,* December 10, 1978, p. SM24.

pp. 220–221 JOHN HUNTER: Kobler, J. (1960). *The Reluctant Surgeon: A Biography of John Hunter.* Garden City, New York: Doubleday and Company, Inc.; Moore, W. (2005). *The Knife Man: Blood, Body Snatching, and the Birth of Modern Surgery.* New York: Broadway Books.

p. 221 WILLING TO TAKE PHYSICAL, SOCIAL, FINANCIAL, AND LEGAL RISKS TO SATISFY THEIR NEED FOR NOVEL AND INTENSE EXPERIENCES FOR THEIR OWN SAKE: Zuckerman, M. (1994). *Behavioral expression and biosocial bases of sensation seeking.* New York: Cambridge University Press.

p. 222   FOOD PREFERENCES: Logue, A. W., & Smith, M. E. (1986).
          Predictors of food preference in adults. *Appetite, 7,* 109–125.;
          Teraṣaki, M., & Imada, S. (1988). Sensation seeking and food
          preferences. *Personality and Individual Differences, 9,* 87–93.

p. 222   DANGERS BEHIND THE WHEEL: Arnett, J., Offer, D., &
          Fine, M. A. (1997). Reckless driving in adolescence: "State" and
          "trait" factors. *Accident Analysis and Prevention, 29,* 57–63.; Burns,
          P. C., & Wilde, G. J. S. (1995). Risk taking in male taxi drivers:
          Relationships among personality, observational data, and drive
          records. *Personality and Individual Differences, 18,* 267–278.; Heino, A.
          (1996). *Risk taking in car driving: Perceptions, individual differences,
          and effects of safety incentives.* University of Groningen, the Nether-
          lands.

p. 223   RISK TAKING AND INTIMACY: Ficher, I. V., Zuckerman, M., &
          Steinberg, M. (1988). Sensation-seeking congruence in couples as a
          determinant of marital adjustment: A partial replication and exten-
          sion. *Journal of Clinical Psychology, 44,* 803–809.; Gaither, G. A., &
          Sellbom, M. (2003). The sexual sensation-seeking scale: Reliability
          and validity within a heterosexual college student sample. *Journal
          of Personality Assessment, 81,* 157–167.; Hoyle, R., Fejfar, M. C., &
          Miller, J. D. (2000). Personality and sexual risk taking: A quantita-
          tive review. *Journal of Personality, 68,* 1203–1231.; McCoul, M. D.,
          & Haslam, N. (2001). Predicting high-risk sexual behaviors in
          heterosexual and homosexual men: The roles of impulsivity and
          sensation seeking. *Personality and Individual Differences, 31,* 1303–
          1310.; Thornquist, M. H., Zuckerman, M., & Exline, R. V.
          (1991). Loving, liking, looking and sensation seeking in unmar-
          ried college couples. *Personality and Individual Differences, 12,*
          1283–1292.

pp. 224–   APPEAL OF CRIMINAL ACTIVITY: Baumeister, R. F., &
225   Campbell, W. K. (1999). The intrinsic appeal of evil: Sadism,
          sensational thrills, and threatened egotism. *Personality and Social
          Psychology Review, 3,* 210–221.; Goma, M., Perez, J., & Torrubia, R.
          (1988). Personality variables in antisocial and prosocial disinhibitory
          behavior. In T. E. Moffitt & S. A. Mednick (Eds.), *Biological contribu-
          tions to crime causation* (pp. 211–222). Dordrecht, The Netherlands:
          Martinus Nijhoff.; Katz, J. (1988). *Seductions of Crime: Moral and Sensual
          Attractions in Doing Evil.* New York: Basic Books, Inc.; Wade, W. C.
          (1987). *The Fiery Cross: The Ku Klux Klan in America.* New York:
          Simon & Schuster.

p. 226 SENSATION SEEKING AT WORK: Zuckerman, M. (2007). *Sensation seeking and risky behavior.* Washington, DC: American Psychological Association.

p. 227 PARENTS AND TEACHERS LEARNING FROM CHILDREN: Dillon, J. J. (2002). The role of the child in adult development. *Journal of Adult Development, 9,* 267–275.

pp. 229– ECCENTRIC SEXUAL INTERESTS: American Psychiatric
233 Association (1994). *Diagnostic and statistical manual of mental disorders* (4th ed.).Washington, DC: American Psychiatric Press.; Cantor, J. M., Blanchard, R., & Barbaree, H. (in press). Sexual disorders. In P. H. Blaney & T. Millon (Eds.), *The Oxford Textbook of Psychopathology.* New York: Oxford University Press.; Kolarsky, A., & Madlafousek, J. (1983). The inverse role of preparatory erotic stimulation in exhibitionists: Phallometric studies. *Archives of Sexual Behavior, 12,* 123–148.; Williams, D. J. (2005). Functions of leisure and recreational activities within a sexual assault cycle: A case study. *Sexual Addiction & Compulsivity, 12,* 295–309.

pp. 233– MORBID CURIOSITY: Abdel-Khalek, A. M., & Maltby, J.
234 (2008). The comparison of predictors of death obsession within two cultures. *Death Studies, 32,* 366–377.; Fabregat, A. A. (2000). Personality and curiosity about TV and film violence in adolescents. *Personality and Individual Differences, 29,* 379–392.; Zuckerman, M., & Little, P. (1985). Personality and curiosity about morbid and sexual events. *Personality and Individual Differences, 7,* 49–56.

Chapter 9
Discovering Meaning and Purpose in Life

p. 235 PEOPLE SAY THAT WE'RE SEARCHING FOR THE MEANING OF LIFE. Campbell, J. (1988). *The Power of Myth.* New York: Doubleday.

p. 235 THE SUN SLICED THROUGH THE WINDSHIELD. Murakami, H. (1993). *Hard-Boiled Wonderland and the End of the World.* New York: Vintage.

p. 239 MEANING MAKERS . . . EXPERIENCE PROFOUND HEALTH AND WELL-BEING: Alim, T. N., Feder, A., Graves, R. E., Wang, Y., Weaver, J., Westphal, M., Alonso, A., Aigbogun, N. U., & Smith, B. W. (2008). Trauma, resilience, and recovery in a high-risk African-American population. *American Journal of Psychiatry,*

165, 1566-1575.; Mascaro, N., & Rosen, D. H. (2005). Existential meaning's role in the enhancement of hope and prevention of depressive symptoms. *Journal of Personality, 73,* 985–1013.; Steger, M. F., Kashdan, T. B., & Oishi, S. (2008). Being good by doing good: Daily eudaimonic activity and well-being. *Journal of Research in Personality, 42,* 22–42.; Urry, H. L., Nitschke, J. B., Dolski, I., Jackson, D. C., Dalton, K. M., Mueller, C. J., Rosenkranz, M. A., Ryff, C. D., Singer, B. H., & Davidson, R. J. (2004). Making a life worth living: Neural correlates of well-being. *Psychological Science, 15,* 367–372.

pp. 239–
240

LEE WHEELER: *The Washington Post,* April, 20, 2007, "On Being."

p. 241

SEARCHING AND EXPLORING REQUIRES ATTENTION, AWARENESS, AND EFFORT: Steger, M. F., Kashdan, T. B., et al. (2008), (p. 102–142).

p. 241

WE REGULARLY EXPERIENCE MEANING: Wong, P. T. P., & Fry, P. S. (1998). *The Human Quest for Meaning: A Handbook of Psychological Research and Clinical Application.* Mahwah, NJ: Lawrence Erlbaum.

p. 241

GANDHI: Buncombe, A. (2007). Gandhi & son: a family tragedy; A darker side of greatness. *The Independent (London).*

p. 242

DEFINING PURPOSE: McAdams, D. (2001). The psychology of life stories. *Review of General Psychology, 5,*100–5,122.; Sheldon, K. M. (2004). *Optimal Human Being: An Integrated Multi-level Perspective.* New Jersey: Erlbaum.

p. 243

HAVING A PURPOSE IS THE FIRM FOUNDATION: Creswell, J. D., Lam, S., Stanton, A. L., Taylor, S. E., Bower, J. E., & Sherman, D. K. (2007). Does self-affirmation, cognitive processing, or discovery of meaning explain cancer-related health benefits of expressive writing? *Personality and Social Psychology Bulletin, 33,* 238–250.; Creswell, J. D., Welch, W. T., Taylor, S. E., Sherman, D. K., Gruenewald, T. L., & Mann, T. (2005). Affirmation of personal values buffers neuroendocrine and psychological stress responses. *Psychological Science, 16,* 846–851.; Lapierre, S., Dube, M., Bouffard, L., & Alain, M. (2007). Addressing suicidal ideations through the realization of meaningful personal goals. *Crisis, 28,* 16–25.; Niemic, C. P., Brown, K. W., Kashdan, T. B., Breen, W. E., Levesque-Bristol, C., & Ryan, R. M. (2008). Being present when facing death: The role of mindfulness in terror management. *Manuscript submitted;* Sheldon, K. M. & Elliot, A. J. (1999). Goal striving, need-satisfaction, and longitudinal well-being: The self-concordance model. *Journal of Personality and Social Psychology, 76,* 482–497.; Sheldon, K. M., Elliot, A. J., Ryan, R. M., Chirkov, V., Kim, Y., Wu, C., Demir, M., & Sun, Z. (2004). Self-concordance and subjective well-being in four cultures. *Journal of Cross-Cultural Psychology, 35,* 209–233.; Sheldon, K. M. & Houser-Marko, L. (2001). Self-concordance, goal-attainment, and the pursuit of happiness: Can there be an upward spiral? *Journal of Personality and Social Psychology, 80,* 152–165.

p. 244   PARENTING AS A PURPOSE: Deci, E. L., & Ryan, R. M. (2000), op. cit.; Sheldon & Elliot (1999), op. cit.

p. 244   THREE PATHS TO PURPOSE IN LIFE: McKnight, P. E., & Kashdan, T. B. (2007), op. cit.

p. 245   WITH A CURIOUS MINDSET, YOU WILL BE MUCH MORE LIKELY TO STUMBLE UPON PURPOSE: Bauer, J. J., McAdams, D. P., & Sakaeda, A. R. (2005), op. cit.; McAdams, D. P., Anyidoho, N. A., Brown, C., Huang, Y. T., Kaplan, B., & Machado, M. A. (2004), op. cit.

pp. 245–   LEARNING PURPOSE FROM OTHERS: Bandura, A. (1977).
246   *Social Learning Theory.* Englewood Cliffs, NJ: Prentice Hall.

pp. 248–   REACTING TO LIFE WHEN OPPORTUNITIES AND
249   CHALLENGES INTERVENE: Affleck, G., & Tennen, H. (1991). The effect of newborn intensive care on parents' psychological well-being. *Children's Health Care, 20,* 6–14.; Affleck, G., Tennen, H., Croog, S., & Levine, S. (1987). Causal attribution, perceived benefits, and morbidity after a heart attack: an 8-year study. *Journal of Consulting and Clinical Psychology, 55,* 29–35.; Affleck, G., Tennen, H., Zautra, A., Urrows, S., Abeles, M., & Karoly, P. (2001). Women's pursuit of personal goals in daily life with fibromyalgia: a value-expectancy analysis. *Journal of Consulting and Clinical Psychology, 69,* 587–596.; Bower, J. E., Meyerowitz, B. E., Desmond, K. A., Bernaards, C. A., Rowland, J. H., & Ganz, P. A. (2005). Perceptions of positive meaning and vulnerability following breast cancer: predictors and outcomes among long-term breast cancer survivors. *Annals of Behavioral Medicine, 29,* 236–245.; Coward, D. D. (1990). The lived experience of self-transcendence in women with advanced breast cancer. *Nursing Science Quarterly, 3,* 162–169.; Coward, D. D. (1994). Meaning and purpose in the lives of persons with aids. *Public Health*

Nursing, 11, 331–336.; Park, C. L., Edmondson, D., Fenster, J. R., & Blank, T. O. (2008). Meaning making and psychological adjustment following cancer: The mediating roles of growth, life meaning, and restored just-world beliefs. *Journal of Consulting and Clinical Psychology, 76,* 863–875.

p. 248　　GROW AND BENEFIT FROM STRUGGLES: Calhoun, L. G. (2003). Posttraumatic growth after war: A study with former refugees and displaced people in Sarajevo. *Journal of Clinical Psychology, 59,* 71–83.; Janoff-Bulman, R. (1992). *Shattered assumptions: Toward a New Psychology of Trauma.* New York: Free Press.; Joseph, S., & Linley, P. A. (2005). Positive adjustment to threatening events: An organismic valuing theory of growth through adversity. *Review of General Psychology, 9,* 262–280.; Tedeschi, R. G., & Calhoun, L. G. (2004). Posttraumatic growth: Conceptual foundations and empirical evidence. *Psychological Inquiry, 15,* 1–18.

p. 248　　RAPE SURVIVORS: Kilpatrick, D. G., Edmunds, C. N., & Seymour, A. K. (1992). Rape in America: A report to the nation. Arlington, VA: National Victims Center and Medical University of South Carolina.; Resnick, H., Kilpatrick, D., Dansky, B., Saunders, B., & Best, C. (1993). Prevalence of civilian trauma and posttraumatic stress disorder in a representative national sample of women. *Journal of Consulting and Clinical Psychology, 61,* 984–991.

pp. 249–　THREE FUNDAMENTAL WAYS THAT PEOPLE MIGHT
250　　GROW FROM DIFFICULTIES: Hobfoll, S. E. (2002). Social and psychological resources and adaptation. *Review of General Psychology, 6,* 307–324.; Hobfoll, S. E., Hall, B. J., Canetti-Nisim, D., Galea, S., Johnson, R. J., & Palmieri, P. (2007). Refining our understanding of traumatic growth in the face of terrorism: Moving from meaning cognitions to doing what is meaningful. *Applied Psychology: An International Review, 56,* 345–366.; Park, Edmondson, Fenster, & Blank (2008), op. cit.

p. 252　　TIME SINCE THE EVENT TOOK PLACE: Helgeson, V. S., Reynolds, K. A., & Tomich, P. L. (2006). A meta-analytic review of benefit finding and growth. *Journal of Consulting and Clinical Psychology, 74,* 797–816.

p. 253　　OPENNESS TO THESE EXPERIENCES . . . EVEN WHEN IT HURTS TO DO SO: Linley, P. A. (2003). Positive adaptation to trauma: Wisdom as both process and outcome. *Journal of Traumatic Stress, 16,* 601–610.; Peterson, C., Park, N., Pole, N., D'Andrea, W., &

Seligman, M. E. P. (2008). Strengths of character and posttraumatic growth. *Journal of Traumatic Stress, 21,* 214–217.; Waugh, C. E., Fredrickson, B. L., & Taylor, S. F. (2008). Adapting to life's slings and arrows: Individual differences in resilience when recovering from an anticipated threat. *Journal of Research in Personality, 42,* 1031–1046.; Waugh, C. E., Wager, T. D., Fredrickson, B. L., Noll, D. C., & Taylor, S. F. (in press). The neural correlates of trait resilience when anticipating and recovering from threat. *Social Cognitive and Affective Neuroscience;* Zoellner, T., & Maercker, A. (2006). Posttraumatic growth and psychotherapy. In L. G. Calhoun & R. G. Tedeschi (Eds.), *Handbook of posttraumatic growth* (pp. 334–354). Mahwah, NJ: Lawrence Erlbaum Associates.; Zoellner, T., & Maercker, A. (2006). Posttraumatic growth in clinical psychology—A critical review and introduction of a two component model. *Clinical Psychology Review, 26,* 626–653.

pp. 253– SEPTEMBER 11, 2001: Fredrickson, B. L., Tugade, M. M., 254   Waugh, C. E., & Larkin, G. (2003). What good are positive emotions in crises?: A prospective study of resilience and emotions following the terrorist attacks on the United States on September 11th, 2001. *Journal of Personality and Social Psychology, 84,* 365–376.; Peterson, C., & Seligman, M. E. P. (2003). Character strengths before and after September 11. *Psychological Science, 14,* 381–384.

p. 255   EXPLORING OUTSIDE THE CONFINES OF OUR LIFE SPACE: Bishop, et al. (2004), op. cit.; Cappeliez, P., & O'Rourke, N. (2002). Personality traits and existential concerns as predictors of the functions of reminiscence in older adults. *The Journals of Gerontology, 57B,* 116–123.; Hidi, S., & Renninger, K. A. (2006), op. cit.; Kashdan, T. B. & Steger, M. F. (2007), op. cit.; McCrae, R. (1993). Openness to experience as a basic dimension of personality. *Imagination, Cognition, and Personality, 13,* 39–55; Silvia, P. (2001), op. cit.

p. 256   TRIAL, ERROR, AND CHANCE PROCESS: Simon, H. A. (1956). Rational choice and the structure of the environment. *Psychological Review, 63,* 129–138.

p. 257   "SATISFICING" SOLUTION: Schwartz, B., Ward, A., Monterosso, J., Lyubomirsky, S., White, K., & Lehman, D. R. (2002). Maximizing versus satisficing: happiness is a matter of choice. *Journal of Personality and Social Psychology, 83,* 1178–1197.

pp. 259– PURPOSEFUL LIVING AND SELF-DOUBT: Dugas, M. J.,
260   Freeston, M. H., & Ladouceur, R. (1997). Intolerance of uncertainty
         and problem orientation in worry. Cognitive Therapy and
         Research, 21, 593–606.; Dugas, M. J., Hedayati, M., Karavidas, A.,
         Buhr, K., Francis, K., & Phillips, N. A. (2005). Intolerance of
         uncertainty and information processing: Evidence of biased recall
         and interpretations. *Cognitive Therapy and Research, 29*, 57–70.;
         Kruglanski, A. W. (2004), op. cit.

p. 261   CURIOSITY IS ONE OF THE MOST COMMONLY EN-
         DORSED STRENGTHS: Park, N., Peterson, C., Seligman, M. E. P.
         (2004), op. cit.; Peterson, C., Ruch, W., Beermann, U., Park, N., &
         Seligman, M. E. P. (2007). Strengths of character, orientations to
         happiness, and life satisfaction. *Journal of Positive Psychology, 2,*
         149–156.; Shimai, S., Otake, K., Park, N., Peterson, C., & Selig-
         man, M. E. P. (2006), op. cit.

Index